WINDS *of* CHANGE

Books by Ramon V. Navaratnam

Winds of Change:
Malaysia's Socioeconomic Transition
from Dr Mahathir Mohamad
to Abdullah Ahmad Badawi (2004)

Malaysia's Socioeconomic Challenges:
Debating Public Policy Issues (2003)

Malaysia's Economic Sustainability:
Confronting New Challenges
Amidst Global Realities (2002)

Malaysia's Economic Recovery:
Policy Reforms for Economic Sustainability (2000)

Healing the Wounded Tiger:
How the Turmoil is Reshaping Malaysia (1999)

Strengthening the Malaysian Economy:
Policy Changes and Reforms (1998)

Managing the Malaysian Economy:
Challenges and Prospects (1997)

WINDS of CHANGE

MALAYSIA'S ECONOMIC TRANSITION FROM DR MAHATHIR MOHAMAD TO ABDULLAH AHMAD BADAWI

RAMON V. NAVARATNAM

Pelanduk
Publications
www.pelanduk.com

Published by
Pelanduk Publications (M) Sdn Bhd
(Co. No. 113307-W)
12 Jalan SS13/3E
Subang Jaya Industrial Estate
47500 Subang Jaya
Selangor Darul Ehsan, Malaysia

Address all correspondence to
Pelanduk Publications (M) Sdn Bhd
P.O. Box 8265, 46785 Kelana Jaya
Selangor Darul Ehsan, Malaysia

Check out our website at *www.pelanduk.com*
e-mail: *rusaone@tm.net.my*

Copyright © 2004 Ramon V. Navaratnam
Design © 2004 Pelanduk Publications (M) Sdn Bhd

Perpustakaan Negara Malaysia   Cataloguing-in-Publication Data

Navaratnam, Ramon V.
     Winds of change: Malaysia's socioeconomic transition from
     Dr Mahathir Mohamad to Abdullah Ahmad Badawi / Ramon
     V. Navaratnam.
     Includes index
     ISBN 967-978-905-5
     1. Malaysia—Economic conditions. 2. Malaysia—Politics and
     government. I. Title.
     330.9595

Printed and bound in Malaysia

To my dear family,
and especially to
my lovely grandchildren,
Suhanya, Sunetra,
Michael Anil and Sarah,
to whom we leave our heritage
for an even better tomorrow.

ABOUT THE AUTHOR

RAMON V. NAVARATNAM is a distinguished former
civil servant and corporate personality. A graduate in
Economics from the University of Malaya in Singapore and
a postgraduate from Harvard University, he was an
economist with the Malaysian Treasury for 27 years, where
he rose to become the deputy secretary-general. During that
time, he also served as alternate director on the Board of
Directors of the World Bank in Washington, D.C. For many
years, he was involved in the preparation of the Malaysian
annual budgets and five-year economic development plans.
He then became the secretary-general of the Ministry of
Transport in 1986.

   After retiring from the civil service in 1989, Tan Sri
Dato' (Dr) Navaratnam was appointed CEO of Bank Buruh
for five years. He is now corporate adviser to the Sunway
Group, executive director of Sunway College, a board
member of the Monash University in Malaysia and the

Monash International Advisory Panel, and a director of the Asian Strategy and Leadership Institute (ASLI).

He served the Malaysian government as vice-chairman of the Malaysian Business Council and as an independent member of the National Economic Consultative Council (1999). He was also on the Board of Directors of the Malaysia External Trade Development Corporation (Matrade) and served on the Board of the Malaysian Industry-Government Group for High Technology (Might). He is also a member of several National Economic Action Council (NEAC) Working Groups. He was a member of the Securities Commission.

He also serves on several voluntary organisations and was deputy president of the Malaysian Association of Private Colleges and Universities (MAPCU) and board member of the Malaysia-U.S. Business Council. He has also recently been appointed to serve as a Commissioner on Suhakam, the Malaysian Human Rights Commission and has been reappointed to the National Unity Panel.

Navaratnam was recently conferred an Honorary Doctorate of Laws by Oxford Brookes University in the United Kingdom in recognition of his many contributions to public service and economic development in Malaysia.

CONTENTS

PREFACE

WINDS OF CHANGE is an analytical commentary on
outstanding issues that emerged in the Malaysian economy
between September 2002 and June 2004. This was a crucial
period in Malaysia's socioeconomic development,
characterised by the transition of the nation's leadership
from Tun Dr Mahathir Mohamad to Dato' Seri Abdullah
Ahmad Badawi. This book also assesses the First 100 Days of
Abdullah's premiership.

During the 22-year period of former Prime Minister Dr
Mahathir's eventful governance, Malaysia's socioeconomic
and political base was considerably strengthened, thus
providing a sound foundation for Abdullah to build upon Dr
Mahathir's momentous albeit controversial era. These have
been exciting times through which we Malaysians have
lived—with some concerns and many impressive
achievements. New challenges emerged domestically and
internationally to test the leadership skills of the government

and the entrepreneurial acumen of Malaysian and foreign businessmen as well as the resilience of all Malaysians.

The critical observation of these challenges and our capacity to overcome them fascinate me tremendously. That is why I felt compelled to write about these historical happenings for posterity to appreciate our hopes and fears and our determination to succeed in building a unique, united Malaysian nation out of so much of our rich racial, religious and cultural diversity.

Consequently, this book—like my previous six books—tracks the developments in Malaysia's response to domestic and international economic challenges as they unfolded and evolved between September 2002 and June 2004.

I hope this book will give a true flavour of our times and help to accurately assess the quality and effectiveness of the management of the challenges and opportunities and the successes or failures that occurred in the Malaysian economy during this transition period.

I can only hope that with God's grace I will be able to continue to contribute my thoughts and analyses of the socioeconomic developments of our times for the benefit of those who come after us.

As in all my previous books, all royalties due to me from the sale of these books will be given to charity. I therefore wish to thank all readers of my books for their warm support and wish them all the best for what they themselves will continue to contribute to our beloved country, Malaysia.

Finally, I would like to thank my tenacious publisher, Dato' Ng Tieh Chuan, and his able assistant, Eric Forbes, as well as my loyal secretary, Haema, for their kind assistance in enabling me to complete this, my seventh book.

Ramon V. Navaratnam
Kuala Lumpur, 2004

CHAPTER I

LIBERALISATION

GLOBALISATION—and liberalisation—is coming fast upon Malaysia. Malaysia's Minister of International Trade and Industry Dato' Seri Rafidah Aziz thankfully announced on September 2, 2002, that Malaysia's "first batch of requests [for market access], for the service sector, will be submitted to the World Trade Organisation (WTO) this month"! They include requests from the engineering consultancy, architecture, accountancy, construction and telecommunications. But where are the requests from the educational, medical, legal and other professional services?

Are they going to be included in the next "batches", and if so, when will it be?

The submission of requests to international trading partners for market opening began on June 30—over two months ago! How much time do we have to make more requests? The "offers" to be made by trading partners, are scheduled to start from March 31, 2003!

Will we be ready to start submitting our offers by then and what offers are we thinking about?

We have to act fast as the target date for the completion of the negotiations is the end of 2004! At this time we thus have only 27 months to finalise our WTO negotiations on the requests, the offers and other complex issues in our service sector.

The Ministry of International Trade and Industry (MITI) believes that the "proactive participation of industry associations is essential"! That is correct.

But it may not be pragmatic to expect our private sector to be enthusiastic about bringing down protective walls to allow more open foreign competition from their more efficient peers!

What then can or should the government do to press the industries to be more forthcoming and internationally competitive? There can be genuine reluctance on the part of our private sector to liberalise at a faster pace.

Industry players such as those in the education services, for instance, may not understand to what extent they themselves would want the industry to open up to foreign competition. There are real national strategic and sensitivities involved.

There is no point in a local industry making requests and offers to our international trading partners if these requests and offers are not going to be approved by the authorities.

Hence MITI will need to take the lead to coordinate with other ministries and guide the industries, on the content, the extent and the timing of making these requests and offers to foreign competitors.

We have to examine the overall national interests from the point of view of the consumers, our national economic plans, our affirmative-action policies and the longer-term competitiveness and sustainability of Malaysia's international trade and investment.

Malaysian industries as well as the public need to be fully appraised of the pros and cons and constraints and limitations, in our future WTO negotiations. We need to know to what extent can we determine the pace of liberalisation. We must learn in greater depth as to what the penalties would be for us if we take more time to open up our markets to foreigners and to liberalise and globalise.

MITI is undoubtedly knowledgeable and well experienced in the art of WTO negotiations, as it has been at it on a continuous basis since the beginning—but that is not necessarily so in the case of all our industries.

Hence we need more open public debate on the questions raised above. After all, it is not only industrial leaders and the government that are involved, but the much larger body of consumers, who have in the end to pay the high price for relatively more protection.

I hope therefore that MITI will provide more explanation to the public on what its plans are, for greater trade liberalisation in the service sector of our economy, as early as possible, as time is of the essence!

MITI could organise more public fora on these important issues since the service sector is vital, as it constitutes more than half our gross domestic product (GDP).

With more public understanding, there will be greater appreciation and stronger general support for more liberalisation and international competition, which are necessary for Malaysia's sustained economic progress.

Hopefully Dato' Seri Abdullah Ahmad Badawi will encourage more public debate on these important trade and investment issues that will be negotiated in the WTO which all Malaysians will have to live with. For this reason alone, they need to be fully appraised of all these WTO developments as they unfold and as they impact on ordinary Malaysians.

Liberalising Too Slowly Internally

But is Malaysia liberalising faster externally and too slowly internally? Is Malaysia globalising too fast with the six original Asean countries, led by the minuscule countries Singapore and Brunei, that have no alternative but to almost completely open their economies to full globalisation?

Liberalisation is good for our economy in gradual stages in the longer term. But to move too fast in the short term can be disruptive. This can happen especially when we are still somewhat cautious about liberalising our economy internally!

There is, I believe, a widening gap between our internal and external economic plans and policies caused by the conflict between globalisation and our affirmative-action policies. If these economic contradictions are not addressed early and coordinated more carefully, there is a real danger that the Malaysian economy will be weakened by rapid globalisation and competition without adequate defenses against external economic attacks!

On the one hand, it is good that MITI is opening up the economy for greater international competition. But, on the other hand, we have internal constraints that inhibit expansion due to our affirmative-action policies that hold back the SMIs that mainly do not enjoy much preferential policies as they are mainly non-*Bumiputera*.

Hence if we exclude our multinational industries, there will be inadequate domestic industrial capacity to compete on even terms, with the manufacturing and service industries of other Asean countries.

4

Specifically, the Industrial Coordination Act 1975 (ICA) and the Foreign Investment Committee (FIC) do curb domestic investment and industrial expansion and dampen competition. Our competitors in Asean and AFTA do not have the same competitive constraints. They will therefore have the competitive advantage over our own businessmen, manufacturers and investors.

The Asean Economic Ministers (AEM) Meeting and Liberalisation

The 34th Asean Economic Ministers (AEM) Meeting in Brunei on September 10, 2002, made considerable progress in hastening the process of liberalisation in trade and investment in Asean, as follows:

1. The six original Asean members will eliminate all import duties by 2010 and the other four new Asean members—Cambodia, Laos, Myanmar and Vietnam (CLMV)—will achieve the same target by 2015.
2. Malaysia agreed to include from the beginning of 2003, all manufactured, processed and unprocessed agricultural products (except rice) into the Common Effective Preferential Tariff (CEPT) Agreement for AFTA.
3. Under the Asean Integration System of Preferences (AISP) now being implemented from the beginning of 2002, Malaysia offered 553 duty-free products to CLMV.
4. At the same time, Asean is planning to set up an Asean Free Trade Area (AFTA) with China! It will cover mostly agricultural products. Can we see Malaysian agricultural products competing effectively with those from China where the farmers

are generally more diligent, cheaper and innovative? The so-called "Early Harvest" package will consist of selected agricultural and other items and could be implemented as early as in the beginning of 2003!

5. Malaysia will also be included in the proposed Asean+3 partnerships with China, Japan and South Korea and the Closer Economic Relationship (CER) with Australia and New Zealand. All these economic relationships could finalise faster than we envisage in the next few years. We have to liberalise and take on the challenges of globalisation, but we must do so carefully, without causing social disruption!

However, the big question remains as to whether Malaysian farmers and manufacturers and businessmen can adequately cope with the huge competitive forces that we will all be up against? *I seriously doubt that most of our businessmen are really ready to face these challenges of liberalisation.*

Implications of Economic Liberalisation

1. What all this means is that there will be a massive inflow of manufactured goods from the cheaper labour industries of most Asean countries. Thus our own small and medium industries (SMIs) will suffer and be forced to close down, raise their competitiveness, or move out to other Asean countries.

2. It will be very difficult for the SMIs to become more efficient fast enough. So we can assume that many SMIs will fade away and unemployment will rise!

3. The 75 unprocessed agricultural products that will be included in the CEPT Scheme will include swine, fowls, live animals, edible vegetables and tropical fruits. So what does this mean for our farmers who produce these products? Will they be able to compete or be swamped by the more competitive and aggressive Thai, Indonesian and Vietnamese farmers? I believe that our farmers will lose out in the competition. Then how will this impact on our policy to reduce and eradicate poverty? Will our social and agrarian stability remain intact?

SMIs and Liberalisation

The Report on the performance of SMIs for 2001-2, based on a survey of 1,009 SMIs, was released by the Small and Medium Industries Development Corporation (SMIDEC) on August 22, 2002. It has raised concern that our SMIs are in trouble over liberalisation. For instance, there are about 48,200 SMIs that comprise 90 per cent of the manufacturing establishments in the country, but produce only 15 per cent of the total manufacturing output. *Their productivity therefore must be detrimentally low! More liberalisation will threaten them!*

Furthermore, SMIs are still in the process of consolidating their operations to address the impact of the Asian financial crisis of 1997. Why are they taking so long and why does the survey not show how many of our SMIs actually went under?

It is discouraging that the SMIs' contribution to the GDP increased by only 1.2 per cent to RM50.8 billion from RM50.2 billion in 2000!

Are the SMIs, which are the backbone of our manufacturing sector and provide a large part of our employment, still so inimical? How much has the government done to help them as compared to the hefty tax incentives given to our huge corporates?

Surprisingly only 44 per cent of the SMIs utilised the assistance provided by the government for training. Yet as Rafidah Aziz rightly pointed out, the SMIs have definite opportunities to supply the large companies and multinationals, provided they have technological capability to meet specified manufacturing standards for costs, quality and delivery. Why is the response of SMIs so poor?

Is it because only 17 per cent of the respondents have adopted Computer-Aided Design (CAD) and only 16.6 per cent use e-commerce solutions in their operations? How can we be competitive enough when the full impact of AFTA sets in soon?

Also just about half (50.7 per cent) of the SMI respondents secured bank loans. Thus a very high proportion (49.3 per cent) had to depend on traditional sources like their savings, and borrowings from families, friends and perhaps even the dreaded loan sharks. Has the banking system been doing enough to help our small industries that really need financial assistance to compete with their counterparts in Asean?

Why are the SMIs not getting more assistance from the government to enable them to be much more internationally competitive? Are there weaknesses due to poor communication and delivery on the part of agencies and the banks? Is it because the mainly Malay officials cannot relate effectively with the mostly Chinese SMIs? We have to rectify these weaknesses urgently to compete internationally.

The situation is made worse when the SMIs are also short of skilled labour. They have to depend on foreign labour to the extent of 31 per cent of their requirements! *Are we then providing more employment and training opportunities to foreign workers who repatriate their savings and foreign exchange back to their countries?*

Unfortunately, after all this time of preparation, and before AFTA comes into operation on January 1, 2003, the majority of our SMIs (79.7 per cent) said they were only at the stage of "preparing to face more competitive business conditions under AFTA"!

When will our SMIs be really prepared to compete with some of our more enterprising Asean businesses—after they export more of their goods and services into Malaysia? Then it will be too late and the adverse impact on our balance of payments, unemployment and our prospects for higher economic growth will be seriously jeopardised!

For all these reasons it is imperative that the authorities should develop more dynamic strategies to enhance the competitiveness of all our SMIs as a matter of high priority, before they are swamped by some of the more efficient AFTA members.

Policy Constraints

The problem is that Malaysia is doing little by way of policy changes and structural and administrative reforms to meet these threats from foreign competition, as indicated below:

1. Land alienation for farmers is slow. Those farmers who can make the best use of land for cultivation of fruit, vegetables, etc., are not given any priority to obtain the necessary arable land.

2. Even when land is approved for cultivation of crops, the process for administrative clearance can be so cumbersome and prone to corruption that our competitiveness becomes blunted.

3. The civil service is still not fully aware of these threats and in most areas there is a lack of a sense of urgency in facing these unprecedented challenges.

4. Some of our leaders also either do not adequately appreciate these strong competitive forces or do not want to publicly express their concerns for the future competitive position of Malaysia. Thus the public too are not aware of the dangers that lie ahead. This can lead to complacency and even to ignorant bliss, as to the real state of our international competitiveness and our future capacity to effectively cope with globalisation.

5. This is why many Malaysian businessmen are lacking in confidence about their future potential and prospects to enhance business. They expect government to undertake major policy changes to meet the challenges of liberalisation—but they see the old mode of operations, the lack of urgency and weak political will, to move at a faster pace to undertake internal socioeconomic reform.

6. Hence, taking into account all these factors, Malaysian businessmen will:
   i. slow down and just wait to see what happens;
   ii. they will just move out to other Asean countries to do business where their efforts will be better appreciated; or
   iii. worse still they will prepare to phase out of business and even rest on their laurels and retire!

Thus Malaysia will have to review and revise many of its present policies that are viewed internally and especially externally as unattractive to more rapid expansion in investment and trade.

When Abdullah Ahmad Badawi takes over as Prime Minister in November 2003, he will need to give higher priority to reviewing all our longstanding trade and investment policies. He has to ensure that we will be moving with the times and not be trapped by traditional thinking, that might be less relevant in today's context of rapid globalisation! *He has to take into account the key concerns of Malaysians and display strong and innovative leadership.*

CHAPTER 2

Key Concerns
of Malaysians

WHILE the major socioeconomic policies have to be constantly reviewed and revised, it is necessary to examine the key concerns of our Malaysian society, so that Malaysian leaders will be able to effectively design the right policies to meet our challenges.

On the 45th anniversary of Malaysia's Independence (*Merdeka*), it is necessary for all Malaysians to be grateful for the peace, stability and progress that we enjoy. It is also a time to look forward to our future prospects as a maturing nation and as a developed country by 2020.

In looking ahead we must examine our concerns with honesty. What indeed are the average Malaysians' concerns?

These issues were discussed at a panel discussion organised by the Malaysian Institute of Management (MIM) in connection with the Tun Hussein Onn Renewal Awards (THORA) Programme on September 28, 2002. The panellists were Dr Chandra Muzaffar of JUST, Sister

Mangalam of the Pure Life Society and myself. The topic was "Key Malaysian Concerns"!

I believe that it is difficult to define the "average Malaysian". It depends on several factors like the age group, the ethnicity, the income level, the opportunities for advancement, the level of education and the capacity to migrate.

Let us take the average middle age group of 35-50 years when most Malaysians would have completed their education, settled in their careers and married with children.

They would generally be asking themselves whether they have attained the right path in their careers, and to what extent they and their children have got a future in Malaysia.

Taking into account the common concerns of most Malaysians, it would be fair to categorise these concerns as follows:

1. **Peace.** Everyone wants to live in a peaceful environment. People everywhere reject insecurity and war.
2. **Employment.** We all need to be employed to enable us to feed our families and to sustain ourselves.
3. **Housing.** A roof over our heads is a basic need, like access to clean water, clothing, electricity, and adequate transport
4. **Education.** It is our birthright and its unavailability can be a serious drawback.
5. **Freedom.** Besides freedom from want, the freedom of speech, assembly and the freedom to practice one's religion and culture, are vital concerns for the well-being of all Malaysians.

The Ethnic Perspective

Since we are a complex, multiracial, multireligious and multicultural country, Malaysian concerns can be best examined from an ethnic perspective.

The Malays (or the *Bumiputeras*) are the majority race in Malaysia. However, they too are not homogeneous in their concerns. The royalty and the political leaders and professionals (Malays and others too) are the privileged class. Their needs and concerns are more than adequately met. Hence they have very few genuine concerns over their well-being and their future.

The educated *Bumiputeras* and politically connected are well placed and enjoy the privileges and benefits of the New Economic Policy (NEP) that is now the NDP. They benefit from promotions in the civil service, scholarships for their children, special equity shares from the 30 per cent of new equity shares reserved for *Bumiputeras*, business licences and especially government contracts. *However, the vast majority of Malays do not have the same access to wealth, mainly because they are not sufficiently educated to take advantage of the government's preferential treatment and because they may not be well connected for these privileges to reach down to them.*

The government is mainly focused on uplifting the welfare of the underprivileged Malays in the lower-income groups. They are the majority and it is they who have given support and continue to provide the ruling United Malays National Organisation (UMNO) with the strong political backing to enable UMNO to form the government continuously since *Merdeka* in 1957!

But many Malays are disillusioned with the NEP. They believe that it is mostly the privileged Malays who have gained disproportionately from the NEP while their lot has not changed significantly since Independence.

This is not fair as Malay poverty has declined sharply due to the vast amounts spent by the government to combat rural poverty. There has been a massive Malay migration into the cities and towns which were primarily occupied by the Chinese and some Indians at the time of *Merdeka*. Unfortunately, however, the NEP has created a "dependency syndrome" especially among the Malays. This "syndrome" has led to high expectations for aid and assistance that has not encouraged self-improvement, self-reliance and enterprise as much as it should. Hence there has been some frustration.

On the other hand, those Malays who have gained considerably from the NEP, in the form of licences and scholarships and government contracts, expect more support. When this is not forthcoming (as there must be limits to aid), there is also resentment against the authorities.

Thus the NEP, which aims to alleviate and eradicate poverty regardless of race, is not always perceived as being evenhanded in its implementation. Non-Malays feel left out!

However, not all Malays have gained from the NEP. At the same time, the poorest among the Chinese and Indians do not necessarily feel that they are given equal treatment as the poorest Malays, in regard to poverty eradication.

This is a major concern of both Malay and non-Malay Malaysians alike. *This is one of the non-Malay dilemmas.* They are Malaysian but sometimes feel they are not treated equitably. This is where Ye Lin-Sheng's *The Chinese Dilemma* (2004) has got it wrong. He claims that the Malaysian Chinese have got a good deal. But his thesis refers to only the higher professional and wealthy business classes to which he belongs.

Part of the problem is that the implementing government agencies are essentially staffed by Malay officials who see it as their duty to help their own kind, even their own kith and kin, with the highest priority! This could be due to the philosophy of their religious *ummah* or Islamic brotherhood.

Chinese Concern Over Malay Dominance

Most Malaysian Chinese (and other races as well) are mainly concerned with the growing tendency of Malay "dominance" over the other Malaysian races. This concept is often described in the Malay language as "*Ketuanan*" or "Bossmanship"!

In politics, the Malays definitely dominate as they are the majority race. Furthermore the political structure in Malaysia is such that the political parties are ethnic-based. So some of the most politically ambitious people in each separate racial group become the leaders. It is not an open competition for the best leaders regardless of race—but some of the toughest leaders *within* each racial group.

This political tradition and practice reinforces the political structure which perpetuates Malay dominance.

Hence the major concern of the Chinese and other races is that they will not be able to compete openly and emerge as natural leaders.

This phenomenon is not confined to the political arena alone. Indeed this political tradition permeates all areas of the economic and social systems at all levels.

The civil service is clearly one important sector that is disproportionately dominated by Malays. Almost all Secretaries-General of the Ministries and Heads of Department are Malay. It is even more pronounced in the military and the police forces. Unlike the private sector, this uneven situation is designed by the government.

Several years ago, when I was a senior officer in the civil service, I stated at a lecture at INTAN (the civil service training centre) that "there is a feeling of alienation among non-Malay civil servants"!

That assessment and feeling has not changed much. Indeed it could have become more pronounced and worsened since the 1970s.

Although the government has recently urged more non-Malays to join the civil service and the uniformed services, the response has been discouraging.

The government has to find out why this is so. It is not (as thought or presented by some authorities), because the civil service pay is unattractive!

An independent survey will probably indicate that the public perception is that it is extremely difficult to get a place in the civil service, and that once you are selected, there will be bias against non-Malays in promotions, good postings, exposure to important responsibilities and other perquisites.

In economics and business it is the same concern over the NEP and now the NDP and the New Vision Policy. The continuing economic policy is to reserve at least 30 per cent of the new equity of public listed companies for the *Bumiputeras* who are mainly Malays. But this percentage is flexible and sometimes a higher percentage is expected by some government agencies!

The policy intention is to strengthen Malay ownership in the business sector as well. Unfortunately, the results here have been less successful. As the Prime Minister Dr Mahathir Mohamad stated in Budget 2003, Malays often sell off their equity shares soon after the shares are provided to them at highly discounted prices.

Many Malays prefer to get rich quickly, without much effort and with little desire to work hard to invest their newfound and easy gains, to generate business growth and expansion.

It is more difficult for influential Chinese businessmen and other non-Malays to obtain new licences to do business and especially to win government contracts on their own merit.

Very often Chinese businessmen and investors have to engage Malay businessmen as partners, to apply and win licences and government contracts. *But often enough these business partners turn out to be "sleeping partners" who enjoy the privileges of partnerships but do not necessarily do more than remain sleeping partners.* This is what is called the *Ali Baba* syndrome, where Ali is the Malay and Baba is the Chinese (or the Indian).

All these factors raise the cost of doing business and squeeze profits. Thus entrepreneurship is often stifled and becomes uncompetitive. This could have serious implications on Malaysia's competitive position with the onslaught of greater international competition from AFTA, China and globalisation. This is a major concern to the smaller Chinese businessmen who wonder how much more they need to sacrifice to satisfy this desire for more government concessions from the more wealthy local businessmen and foreign investors.

Also, many Malay businessmen who were thought to be "entrepreneurial" were identified by the government to become big businessmen. This is only right and proper, so as to create a Malay industrial entrepreneurial class.

Unfortunately again, many, if not most, of these Malay businessmen let down the government, the Malay community and the country. They messed up the management of these large conglomerates that were built up with government and Malaysian (mainly Chinese) taxpayers' support!

Another great concern of the Chinese community is: what will happen to their language and culture in Malaysia?

Although the Chinese have the enviable privilege of having their own Chinese primary (national type) schools and even several large private Chinese secondary schools, where Chinese is the medium of instruction, they are not sufficiently and actively encouraged by the government.

The Chinese language is not taught in the government's national schools, hence most Chinese send their children to Chinese primary schools. Thus polarisation starts. Chinese students also feel alienated when they are deprived of the limited places in government universities, where preference is given to *Bumiputera* students.

Indian Concerns

The Indian concerns are divided like the other races between the wealthy and the poor. About half the Indian population of about 8 per cent or 1.8 million are outside the big towns and in the rubber and oil palm estates. The bulk are in the lower income groups, but there is a wealthy upper class made up mainly of professionals and semi-professionals who are much more comfortable than the estate workers and the large numbers of the urban poor.

The professional Indians are concerned about their future in terms of whether their children will have the same opportunities for advancement as their parents. Where there is doubt they increasingly think of sending their children abroad for a good education and also often consider emigration for their children if not for themselves!

However, the low-income Indians are more concerned with Tamil education and employment opportunities. The government employs them as a last resort. The Chinese—as well as the Malays—in the business sector also prefer to employ their own kind for political reasons. Thus the Indians feel marginalised—unless they are well qualified and at the technical and professional levels.

Those Indians who are semi-educated therefore have limited access to the employment market. Many Indians then turn to gangsterism and other criminal activities.

Part of the problem is that the government has not given much attention to the urban poor, a significant number of whom are Indians. Anti-poverty programmes have been successful in the rural areas which are basically Malay, but have been inadequate in the urban areas in the past.

The other cause of Indian poverty and relative neglect is that they do not have as much political and electoral clout as the Malays and the Chinese. Dato' Seri S. Samy Vellu, the longstanding leader of the Malaysian Indian Congress (MIC), has been forceful in his pleas for more priority to be given to the Indian community. However, the response has been chequered and is not sustained.

As long as there are Tamil schools that do not get sufficient financial support from the government, they won't be able to provide quality education for the Indian children, especially in the plantation estate sector.

The concerns of Malaysians and the individual ethnic groups are deep and wide. The only way to help assuage these concerns is to ensure that the economy keeps growing at a steady pace and that the growth is more equitably shared among all Malaysians. This has been the challenge since *Merdeka* and will continue to be the main concern of all Malaysians for a long time to come. The questions of equity and justice are vital concerns for they will determine the stability and sustainability of the Malaysian economy and the sovereignty of the nation—as national unity is at stake!

The Concerns of the Smaller Minorities

Very often we tend to overlook the fact that there are other important minorities in Malaysia, especially in Sabah and Sarawak. They are the Ibans, the Dayaks, the Muruts and other ethnic groups, including the *Orang Asli* (Aborigines). They constitute the non-Malay *Bumiputeras* (or "sons of the soil"). They are the original settlers in Malaysia, large numbers of whom are non-Muslims. These minorities are not as assertive and articulate as the other larger ethnic groups. Thus they are often not adequately taken into account in the planning of the mainstream socioeconomic development of the country.

Indeed, much more attention would need to be given to these minority groups and the large numbers of other marginalised groups, regardless of race and religion—if we are to achieve the Vision 2020 goals of strong national unity and equity in our country.

All these concerns of thinking Malaysians cut right across race and religion and are primarily the preoccupation of the mass of the Malaysians. The rich and the privileged may be less concerned with these issues of equity and justice.

Abdullah Ahmad Badawi, as Prime Minister since November 2003, will therefore need to give these vital concerns more priority as they would have a major impact on human welfare and Malaysia's overriding aim to create a strong and united Malaysian nation! The Malaysian civil service has a major role in helping the new Prime Minister to build national unity. But can the civil service deliver?

Changes in the Civil Service
for Nation-Building

The Malaysian Administrative Modernisation and Management Planning Unit (MAMPU) has contributed greatly to the development of management and administration capacity in the public sector in the last 25 years since its establishment.

Former Prime Minister Dr Mahathir Mohamad has rightly pointed out that we can be proud to have the best civil service in the developing world. But we are now an advanced developing country—well on our way to developed-nation status by 2020! *We should therefore compare our standards to that of developed countries!*

So what is the future role of MAMPU in improving the quality of the civil service to enhance its delivery system? What changes do we face and what are the critical factors for nation-building by the civil service?

Critical Malaysian Concerns

1. *The economy has developed and modernised faster than expected—but the public service has been strained and unable to cope adequately.* The evidence is in the poor counter services and the many complaints to the Public Complaints Bureau (PBC), which indicate that:
   i. The Land Offices provide some of the worst services!
   ii. Public perception of the efficiency of the civil service is perhaps at its lowest point.
2. Most ministers are well qualified and competent and have become more dominant, but the civil service cannot cope with them.
3. The public expectations of the civil service are much higher as the public has become far more sophisticated. They are more aware that they pay the taxes for civil servants' salaries and thus expect much better services!

4. The public service is attracting and retaining relatively less qualified, less competent and poorly motivated civil servants!

5. The management and personnel systems in the government have not been sufficiently modernised, to meet the new challenges of "Best Management Practices"!

6. Globalisation will bring about a wide range of changes rapidly. The main challenge will be greater pressure to compete internationally. This will entail the need to increase competition internally.

7. Thus the NEP will have to encourage the poor and the privileged groups to move away from, rather than to continue to seek protection. We have to teach our children and grandchildren that the government can only do so much, and that they have to secure their own future by becoming more competitive and competent as early as possible!

8. Costs of production will be challenged severely—corruption, delays and inefficiencies will take their toll on the economy and on attracting foreign investment.

Public Perceptions of the Civil Service

The following are some of the public perceptions of the civil service in Malaysia.

1. *The civil service is comprised mainly of one race and is therefore not sufficiently empathetic.*

2. Malaysia Incorporated works mostly for the select few and does not permeate the whole system or society. This is where some observers make wrong analyses of the Malaysia Inc. practices.

3. Corruption is entrenched in the system and is becoming worse—and the NEP is being eroded because of it, as the poor Malays and others suffer the most.
4. Public investment is too costly, overpriced and often wasteful because of corruption.
5. There is little political will to raise efficiency, transparency and accountability, despite all the well-intentioned statements at the highest levels. This is because the authorities appear to be reluctant to take consistent and sustained tough action against corruption especially at high levels.

Solutions for the Civil Service

So, how do we go about tackling these negative public perceptions of the Malaysian civil service.

1. Recruit better qualified candidates by making the civil service more attractive.
2. Reward productivity more than seniority—more seriously. Be tougher on disciplinary measures and make it easier to discipline unproductive staff, instead of treating them too gently.
3. Provide more training that is more relevant and more efficiency-oriented:
   i. Career prospects must be made more exciting and fulfilling.
   ii. The civil service should always be independent, nonpolitical and professional as well as more representative of Malaysia's multiracial society.
   iii. The resistance to recruit more multiracial staff must be overcome.

We need to restructure the public services. Then we would be seriously recognising the internal and external changes taking place. We will then be prepared to face the challenges, with greater confidence, in building our nation, towards achieving industrial-nation status by 2020!

But the world economic prospects will determine our domestic economic environment and outlook.

CHAPTER 3

The Economic Environment in 2003

AT the dawn of 2003, the global and Malaysian economy continues to be weighed down by the terrible prospects of a U.S. war in Iraq.

So what is the economic outlook for the Malaysian economy in 2003?

The business climate and the business mood is not positive. Indeed Malaysian business confidence is low, with all this continuing uncertainty in the international economy.

It is therefore encouraging that the Second Minister of Finance Dr Jamaluddin Mohd Jarjis stated that he will soon be announcing new measures to fine tune economic policies and the strategy. This is to take into account the changed environment after the September 11 terrorist attacks in the U.S. in 2001 and the October 2002 Bali bombing in Indonesia.

The Executive Director of the NEAC, Dato' Mustapa Mohamed, has also stated that it is too early to talk about the details of the specific measures to sustain economic growth. He is right. Any new package must be carefully thought through as it will have a major positive impact on the economy.

As Prime Minister Dr Mahathir Mohamad firmly said in his 2003 New Year message, Malaysians must have "confidence" in themselves and in the government strategies to solve problems and develop the country.

It is therefore vital that the new economic policies will strongly boost public confidence and not fall short of public expectations! We should seize this opportunity to arrest any declining confidence.

But basically there have to be new government economic policies that will enable the economy to be more transparent, liberal and competitive—even if the government does not have the financial capacity to continue to pump-prime the economy.

At the same time, the corporate ownership requirements need to be more macro- rather than micro-managed so as to encourage Malaysian businesses to expand.

The MIER National Economic Outlook 2003 Conference reported that the GDP would accelerate to 5.7 per cent in 2003 and 6.3 per cent in 2004! These are bold estimates.

Credibility in economic estimates can decline, unless we give all the scenarios and not just make optimistic assumptions. The business community worldwide is pessimistic, given the serious uncertainties regarding the weak global economy and a possible war in Iraq and the explosive situation in the Israel—Palestinian conflict!

The "reality check" that the government intends to introduce must accurately and clearly inform the Malaysian public of the real state of the economy. It is important that

Malaysians fully understand the serious challenges facing our economy at this time of international economic uncertainty.

Although our economic fundamentals are still strong, they can be severely strained, if the Americans attack Iraq or the Western economies and Japan continue to be weak and the whole world sinks into a deflationary spiral!

The U.S. Federal Reserve Chairman Alan Greenspan told the Economic Club of New York on December 20 that the "the U.S. is nowhere close to sliding into a pernicious deflation", like the deflation cycle in Japan!

Greenspan thinks that U.S. monetary policies could arrest any deflation in the economy. But that is not true as the U.S. interest rates have been cut back persistently to a four-decade low of a mere 1.25 per cent. How much lower can it go?

Such low interest rates provide no incentive to save. A vicious Savings-Investment Gap can emerge which will further weaken the U.S. economy and indeed cause the very deflation trap that the U.S. is trying to avoid!

What the U.S. and the world economy really need is for the U.S. to cut back severely on its vast unproductive defence expenditures, which the U.S. war-mongers and armament industries are clamouring for.

Instead the U.S. should invest more in civil works and programmes that will directly benefit the U.S. economy and the American people—and not the "industrial-military complex and the Washington Consensus"!

Christmas shopping in December 2002 in the U.S. has been the weakest on record! Spending on durable goods (over the last three years) as well as spending on plant and machinery (which sustains economic recovery and growth) have actually declined! So where is the hope for the U.S. economy to strengthen in the short term?

However, massive defense spending, on the pretext of destroying Iraq's capacity to produce weapons of mass destruction (WMDs), could quickly boost the sputtering economic engine of the U.S.! This indeed is what the Washington Consensus may want to achieve—to stimulate the economy and to gain a stronger control of Iraq's large petroleum reserves!

Prime Minister Dr Mahathir stated on Christmas Day 2002 that Malaysia will be affected less than before from any war in Iraq, because we have diversified our exports away from the U.S. in recent years.

This is a fair assessment except that the U.S. still imports about 20 per cent of our exports. Thus any U.S. recession could still have a major adverse impact on our economy.

It would therefore be more important now, to exert even stronger efforts to diversify our exports within the more reliable markets of Asia. With the vast China markets and the industrial countries of Japan, South Korea and Taiwan, we could import more of our heavy plant and machinery from these industrial countries, rather than depend too much on the strong but unstable industrial-military-based U.S. economy?

Already the U.S. government has asked Congress to raise the debt limit, from the present US$6.4 trillion. This would strain the U.S. budget deficit even further!

Where is the U.S. economy going? Can we trust its capacity to manage the world's largest economy—efficiently?

Similarly in Japan, Finance Minister Masajuro Shiokawa has expressed regret that nearly half its Budget had to rely on new bond issues!

This heavy debt financing has been made to counter the severe drop of 10.7 per cent to 41.8 trillion yen, in the Budget revenues, a shortfall of 2.54 trillion yen from the earlier Budget estimates. This slack in revenue indeed reflects the deflation in the Japanese economy. On the other hand, public works expenditure has been reduced by 3.7 per cent to 8.91 trillion yen.

The latest estimates of the Organisation for Economic Cooperation and Development (OECD) indicate that the U.S. would expand by only 2.6 per cent, Europe by just 1.8 per cent and Japan by a mere 0.8 per cent in 2003!

These dull economic prospects in the industrial world, together with the increased competitive challenges from China and AFTA, show that Malaysia's economic outlook, cannot be expected to be bright in 2003!

China and South Korea, with their projected growth rates of 7.5 per cent and 5.4 per cent respectively appear to be the few engines of growth for the time being! Hence the case for Malaysia to "Look East" is more relevant now, than ever before. We find that we cannot rely on the U.S. and Europe as they get embroiled in wars of destruction and imperialistic pursuits for hegemony, that are inimical to their own and the world economies!

That is why it is imperative for Malaysia to face up to these dire developments. We can only be better prepared to meet these serious challenges if we adopt strong new economic policies—soon!

Restructuring the Malaysian Economy

What then can we do about these poor economic prospects?

Simply put, we have to restructure the Malaysian economy to meet these major and new challenges as a matter of priority—before it is too late!

For instance, as the only guest of the Malaysian International Chamber of Commerce and Industry (MICCI) General Committee Luncheon recently, I was impressed with the frank views of their President Jose Lopes. He said that more foreign investment will flow into Malaysia, if we make them even more welcome to open more regional headquarters in Malaysia.

Then we need not be unduly concerned with the growing competition from China and AFTA, as we could be exporting to these huge markets rather than losing business to them! So why do we not do even more to attract regional headquarters into Malaysia?

There are still many complaints of a great deal of frustration over the many persisting bureaucratic hitches and delays.

Although foreign investors recognise Malaysia's good intentions at the top policy levels to improve the business environment, they nevertheless continue to face many difficulties on the ground, for timely approvals of one kind or another!

Why can't we resolve these long outstanding civil service problems fast enough? Is there a lack of will or are we too tolerant of inefficiencies in the whole administrative machinery, especially at the lower levels of implementation.

Are some civil servants waiting for rewards before they take prompt action? Or have we just given up wanting to change for the better, because we are now suffering from "civil service fatigue"!

Whatever the reasons, I sincerely hope that any new policy measures will give top priority to breaking the back of these nagging bureaucratic problems—once and for all.

Otherwise, any so-called new measures will fall on deaf ears and sound like a broken record!

All the good policies and the best efforts of the government will come to naught, unless the public service has the will or is forced to deliver!

The government therefore has the immediate responsibility to stop the growing rot in its delivery system—if we are to overcome our economic challenges.

The business sector will watch and wait—but it cannot afford to be too patient as it will lose out to its competitors and investors will then move away from Malaysia! So let us act fast or regret at leisure.

I spoke at the Panel organised by the 25th Anniversary of the Malaysian Administrative and Modernisation Management Planning Unit (MAMPU) in December 2002.

I could sense that the Chief Secretary to the Government of Malaysia Tan Sri Samsudin Osman and most of his senior officers are highly committed to improving the civil service. However, there is a major problem in ensuring that this same spirit of service to "God, King and Country" is shared down the line.

There appears to be a lack of a sense of urgency and competence that worsens as we go down the line of seniority in the public service.

Thus the civil service may become isolationist, too comfortable and even complacent in the ivory towers of Putrajaya—far removed from the realities and problems of the grassroots all over the country. There is indeed the worry that the civil service may be beginning to lose its soul to serve with dedication and diligence! The evidence of inefficiency is apparent when we see the long queues at some counter service departments.

I do not understand why we cannot solve this perennial problem by posting more and better trained staff or impose more discipline, to ensure better counter services to the patient and taxpaying public?

The crux of the matter of poor counter services (and other services like those in the Land Offices!), is really the weak "value system" and the "bad attitude" of many of our civil servants.

This could be overcome to a large extent by recruiting better qualified and multiracial staff and providing more training. In the end, however, the raw material for training must be good or at least suitable!

But the quality of our recruits depends on the quality of the graduates that the education system turns out from our school system!

The National School System

Thus Prime Minister Dr Mahathir's pertinent statement that he made in an exclusive interview with the *New Straits Times* and *Berita Harian* on Christmas Day 2002 is profound.

He said that "the national school system has been hijacked by obscurantists interested in Islamic practices that emphasise form over substance"!

This is one of his most critical remarks that he has made about the present education system. It must be taken very seriously and followed up with remedial measures!

Hence it is significant that the Prime Minister himself will be heading the Committee to reform the education system! Then all avenues to "hijack" the system by deviationists and extremists should be blocked, without leaving any loop holes for any negative policies in the future! Education of our children and posterity is too precious to be left to mere politics and manipulative influences and practices!

Reforming the Public Service

The same reformist approach would need to be adopted to the civil service and the armed forces and the police services.

They are now not sufficiently multiracial, to reflect the ethnic composition of our multiracial society.

To that extent, the credibility of the public services is often called to question. Because they are dominated by one ethnic group, the other ethnic groups are not well disposed to the public and uniformed services. They are often unduly and sometimes unfairly critical of the efficiency of the civil service, particularly at the lower levels.

The solution is to step up efforts to make the public services much more multiracial and efficiency oriented, without being overly protective of the "civil service community"!

Of course, the usual claim is that the Chinese and the Indians and others are reluctant to join the civil service and the armed forces. But it begs the question as to why this has been the case for so long?

The answer is largely because many civil servants have also abused their responsibilities to recruit and promote other ethnic groups into the public service and while they are serving in the public services!

Non-Malay staff will tell you repeatedly that for most of them, their chances for recruitment and promotion are far less than their Malay counterparts.

Thus frustration grows in the civil service. Then the perception seeps through the whole system—and outside the public domain—that the non-Malays have little prospects in the government services!

Hence it would be useful if the government now adopts some kind of affirmative-action policy on recruitment and promotion of non-Malays in the public service! This will ensure that the Malaysian public services and the armed forces as well as the police force are made welcome and attractive to the non-Malays.

The former quota system could be applied in reverse, to insist on greater "balance" in the whole public service. This will encourage stronger national unity, competitiveness and efficiency in the government services.

This important reform will strengthen our economic resilience, our sustainability and also motivate the people and the nation as a whole, to achieve industrial-nation status by 2020!

Therefore, if Malaysia is to progress at a faster pace to be more internationally competitive, we have to introduce more innovative initiatives and aggressive administrative reforms, in our management of the Malaysian economy.

FTA with Japan

One important new economic initiative is the announcement from Prime Minister Dr Mahathir Mohamad in Tokyo on December 13, 2002, that Malaysia and Japan will start discussing a broad-based bilateral Free Trade Area (FTA).

We have no alternative but to purposefully pursue this line of global expansion of trade and investment, now that Singapore has broken the tradition of Asean cooperation and cohesion.

Singapore was the first to sign FTAs outside Asean and now we have Thailand. Hence my earlier hopes for the advantages of having FTAs between Asean as a whole and other major economies like Japan, the U.S., the European Union (E.U.), China and South Korea and Australia—have diminished.

This is because some Asean countries like Singapore did not see the advantages of Asean unity and collective Asean negotiations with these major economies, to enhance

Asean's bargaining position. So much then for Asean's future cohesion and strength!

Asean countries will from now on, each one go more and more on their own way—and the Asean spirit will gradually fade away. I hope this trend will not develop at the expense of Asean unity—thanks to the Singapore's selfish initiative on establishing FTAs with so many industrial countries, regardless of Asean's collective interests!

Malaysia's economic interests could further worsen if world economy moves into a deflationary spiral, especially in Japan and the U.S.

Thus new economic policies are being adopted to face these serious challenges. But all these economic initiatives being planned in the U.S., Europe, Japan and even in Malaysia, to counter deflation, can be adversely affected—if the U.S. and its allies attack Iraq, on some pretext or other after January 27, 2003!

War in Iraq after January 27, 2003?

January 27 is the deadline for U.N. Chief Inspector Hans Blix to give his second report to the U.N. Security Council after his first report on December 8, 2002.

Already the U.S. Ambassador to the U.N. Security Council John D. Negroponte has commented that the Iraqi Arms Declaration made on December 7 to the U.N., is in "material breach", which means that it can find the excuse it wants to attack Iraq unilaterally—for its rich oil reserves!

No other country, including America's staunch ally Britain, has taken that arbitrary stand, before fully studying the Iraqi Declaration which has not even been entirely translated from its Arabic portions!

However, Blix has stated that "An opportunity was missed in the Declaration to give a lot of evidence". But he quickly admitted that he had found no evidence so far to prove that Iraq still has weapons of mass destruction!

Hence the U.S. insists on getting some Iraqi scientists and their families out of Iraq, so that they can be interrogated for information on hidden weapons of mass destruction that the U.S. and the U.K. jointly accuse Iraq of hiding!

The U.S. also wants to get some "Iraqi defectors" to spill the beans—but that could mean that the U.S. would try to bribe the Iraqi scientists with promises of wealth and political asylum and even U.S. citizenship to get the story the U.S. wants! Blix replied that he is "not going to serve as a defection agency"!

Blix has emphasised that he is not going to hijack Iraqi scientists out of Iraq, because that is not part of his job. In any case, all the other 14 U.N. Security Council members think the idea of getting Iraqi scientists out of Iraq is preposterous. But the U.S. may not care!

Instead Blix has asked the U.S. and the U.K. to cooperate fully with the U.N. Inspectorate by sharing intelligence on Iraqi weapons of mass destruction—if they have the secret information. After some initial resistance, "for security reasons", the U.S. has now given in to Blix's appeal and its intelligence will now be given. But what will be its quality?

Hence we have to wait for January 27, 2003, to see if the U.S. and the U.K. are right in their accusations against Iraq or whether it has been a great hoax and an excuse to attack Iraq to capture its vast oil resources? At this time of uncertainty we cannot see the outcome of the U.S.'s imperialistic intentions—but only time will tell!

37

In the meantime, we cannot just sit and wait! We have to look after our own house and put it in better order to face up to the rapidly changing challenges.

Smart Partnership for Industrial Peace

Therefore we have to build our economic resilience by ourselves, regardless of whether individual Asean countries will work together with us or whether war breaks out in the Middle East!

The government, the business sector and labour should thus work more closely together, in the spirit of Malaysia Incorporated, to sustain Malaysia's developing economy.

Thus it is vital to have industrial peace. It is unfortunate that there appears to be increasingly strained relations between the employers associations and the trade unions.

The President of the Malaysian Trades Union Congress (MTUC) Senator Zainal Rampak has said that the MTUC is prepared to go on the war path if the Malaysian Employers Federation (MEF) pushes the government to adopt its "productivity-linked wage system"! He reportedly stated that industrial unrest will occur if the "guidelines on wage reform", accepted by the Tripartite Agreement among the MTUC and the MEF on May 27, 1996, is used as a "licence for employers to bully its workers"!

The President of the MEF, Jafar Carrim, had earlier claimed that the MTUC had made "an about-turn" after agreeing to the Tripartite Agreement.

But Zainal Rampak is right. Guidelines on wage reforms are useful but are not rigid criteria for determining productivity, but only guidelines. They need to be carefully studied collectively, by all parties to arrive at fair and proper principles that would clearly define how to measure and reward productivity.

After all, not all work can be quantified and reliably measured. For instance, the output of employers and managers is the most difficult to assess quantitatively! Thus productivity measurement is a sensitive issue by its very nature.

The MTUC should not be made to feel that the responsibility for determining productivity will lie sorely with the employers or management alone.

Then the fear will naturally arise as to whether union leaders will be penalised for bargaining with management over productivity measurements.

In fact, labour leaders who work hard to contribute to industrial peace should be recognised for their productivity, in providing a conducive industrial environment!

Therefore the concept of "Strategic or Smart Partnership" has to be adopted by the government, employers and labour, so that genuine consensus will be developed on a National Wage Policy that will be based on the productivity criteria, one that is professional, fair and reasonable to all parties.

The MTUC has not opposed the need to increase productivity as it realises that labour will also lose out if Malaysia becomes more uncompetitive, given the challenges now increasingly posed by globalisation and low wage countries, like some Asean countries and especially China. What the MTUC wants is fair play in the implementation of the Tripartite Agreement on the "Guidelines on Wage Reform". There is therefore no need for any party to accuse anyone of "about turns", as we are all in the same boat, in our journey towards achieving Vision 2020.

I hope the government takes a stronger lead in getting all parties to work more closely together to reform the present wage system, for the benefit of all partners in Malaysia's dynamic development process. We cannot afford unproductive dissension and disunity, particularly at this time of increasing international competition.

The government has not as yet introduced a "productivity-based wage structure" in the economy. The trade unions and the private-sector employers do not appear to want to adopt such a system that can benefit the whole economy.

This matter may well be another major issue that will be left for Abdullah Ahmad Badawi to tackle and introduce, when he is in a position to do so as the Prime Minister.

Wage reforms are necessary to make our economy much more competitive. International competition is becoming more severe and this requires both labour and management to be more highly competent.

And as if to add fuel to fire, the U.S. is pushing its own self-interests so hard, that it could erode the economic competitiveness and prospects of developing countries like Malaysia!

A Tariff-Free World to Benefit Mainly the Rich

For instance, U.S. Ambassador Marie T. Huhtala's wrote an interesting article in *The Star* entitled "Towards a Tariff-free World by 2015", on December 10, 2002. She mentioned that the U.S. had made a bold proposal two weeks before to the WTO, "to eliminate all tariffs on manufactured products by the year 2015". She explained that "we are driven by the belief that trade liberalisation benefits all"!

Therein lies the major flaw in U.S. trade policies towards the developing countries. It is well known that in fact the rich and powerful industrial countries will benefit the most, while the poor Third World developing countries, will gain the least from the elimination of tariffs on agricultural products and manufactures and unduly rapid trade liberalisation.

Developing countries are mostly agricultural countries, whereas the rich industrial countries are mainly advanced manufacturing- and service-oriented economies.

But the industrial countries have had tariff walls for manufactures since the industrial revolution and colonial times, for hundreds of years. But now they prevent the developing countries from having protective tariffs.

Now the U.S., as the richest and most powerful country in the world, is ready to propose zero tariffs on manufactures and wants struggling developing countries to do the same in just 13 years, that is, by 2015!

Where is the logic, the economic justice and equity? No wonder there is so much opposition among fair-minded people even in the U.S. and other industrial countries against this self-centred brand of oppressive globalisation.

If the U.S. and some other industrial countries are really sincere in reducing poverty in the developing world, why do they not reduce their vast agricultural subsidies?

It was only last week that Eveline Herfkens, the U.N.'s executive coordinator for the Millennium Development Goals (MDGs) Campaign, said that, *"the rich world's efforts to protect its food producers were keeping poor countries poor"*! (At the U.N. Millennium Summit in September 2000, world leaders placed development at the heart of the global agenda by adopting the Millennium Development Goals, which set clear targets for reducing poverty, hunger, disease, illiteracy, environmental degradation and discrimination against women by 2015.)

No wonder there is mounting anger against the U.S. and the rich countries, that is often expressed in the form of terrorism and a clash of civilisations!

The Rich Countries Should Stimulate the World Economy

What the U.S. and some other rich countries should do, is firstly to eliminate all agricultural subsidies that stifle the food production in many poor developing countries, and secondly, to introduce more gradual and differential target dates for the elimination of tariffs on manufactured goods and the service industries, in the developing countries, according to their state of development.

I hope the affable Ambassador Huhtala could advise Washington to fully take into account the legitimate aspirations and the well-being of developing countries as well, when they formulate U.S. trade and investment policies. They should not sell us the line of the Washington Consensus wholesale as it undermines our development and increases the U.S. dominance over the developing world.

That would be one important way for the U.S. and some rich and powerful industrial countries, to make more friends and not war, and to promote greater world peace and prosperity, for all countries in the future!

Finally, the prospects for reduced international terrorism and world peace and progress will largely depend on whether these leading world powers adopt enlightened policies and practices that are fair and just to the whole world—and not policies that are designed mainly for their continuing world hegemony.

We will have to constantly monitor international developments that will affect our small economy and face up to the challenges confronting the Malaysian economy.

CHAPTER 4

How is Malaysia
Facing its Challenges?

REVIEWING the Malaysian economy in February 2003, it
is considered that the economy needs a boost, given the
international economic uncertainty, arising from the
looming U.S. war against Iraq.

That is why the much-awaited National Economic
Action Council (NEAC) Stimulus Package, expected to be
released at the end of March 2003, is critical.

*The Second Finance Minister Jamaluddin Mohd Jargis has
announced that the Stimulus Package will also include
contingency plans to counter the adverse effects of a possible
U.S.-led invasion of Iraq.*

He said that the government has set up 10 committees to:

1. Increase national competitiveness;
2. Improve the country's delivery system;
3. Promote new sources of growth;
4. Strengthen the capital market;

5. Encourage domestic investment;
6. Improve the revenue collection system;
7. Monitor the macroeconomic policies;
8. Encourage investments and tourism;
9. Spur the agricultural sector; and
10. Improve security, especially in the transport chain.

This is great news to Malaysians of all walks of life, and especially to businessmen whose confidence in the economy has been shaky because of the prospective war on Iraq.

Indeed this could well be a wonderful opportunity to restructure the whole economy—to make it much more productive and internationally competitive.

The Stimulus Package therefore has to be substantive and not superficial. It has to be far more than just a fiscal stimulus. It has to be structural in order to be significant and credible!

In the near term a fiscal package will help boost economic growth and make us "feel good"!

But for the longer term, the Stimulus Package has to keep the growth momentum going, by its bold policies, not only to restructure but to change the direction of the whole Malaysian economy!

This is will require more than mere pump priming. It will need more basic policy changes in the economy.

Proposed New Goals

In regard to the 10 committees that have been set up by the government to design the Stimulus Package, I hope that the government will include the following goals in the stimulative strategy.

1. Reduce protection and liberalise the economy at a faster pace with target dates for the further liberalisation of different economic sectors, particularly in the service sector.
2. Ensure that the civil service becomes more efficient by using more of the carrot-and-stick approach. The civil service must be worthy of its wages that are paid by the taxpayers. We cannot avoid stronger disciplinary measures just because the civil service can muster about three million voters (if we take husbands, wives and relatives)! We also have to realise that the majority of the civil servants are honest and hardworking. Therefore, the minority laggards should not be allowed to give the majority a bad name!
3. New incentives should be given to develop new sources of growth as well as to actively promote existing growth areas, like education, housing and health and many other services that have great export potential, but which presently feel stifled.

    It is hoped that the NEAC will also introduce and supervise the effective implementation of the stimulative policies for the major service industries like Education, which are so vital for quality socioeconomic growth.
4. The capital market can expand considerably if the rules and regulations that govern them are more expeditiously administered and the regulatory environment is made more pragmatic, development oriented and friendly. We cannot be too strict and inadvertently kill the goose that lays the golden eggs!

5. To increase domestic investment, the longstanding Industrial Coordination Act and the Foreign Investment Committee constraints should be removed or the guidelines should be made more relaxed, if we really want to encourage more domestic and foreign investment as well.

6. **Revenue collection can be further improved.** But the professional skills of the Inland Revenue Board (IRB) staff have to be enhanced by providing more sophisticated training. More private-sector experts who from personal experience know how the large multinationals and other businesses use "transfer pricing" and "parallel accounting" techniques have to be hired by the IRB to collect more revenue!

7. **Macroeconomic policies have to be significantly changed.** *The NEP and its implementation have to be considerably modified, to improve the cost-benefits to the nation as a whole. The creation of a rentier class of Bumiputera businessmen has to cease.*

   We must be more meritocratic and competitive at home to be able to compete more effectively internationally, with the advent of greater globalisation. Although our economy is diversified it is still quite vulnerable. Hence we have to make bold moves to raise productivity and efficiency!

8. **Tourism can be improved if more international airlines are attracted into Malaysia.** Our tour packages and the whole tourist industry itself, has to be more tourist friendly and efficient. Presently we do not have a sufficiently strong and professional tourist service culture as compared to Thailand, Singapore and other neighbouring countries.

46

Although the tourist industry provided RM43 billion or about 12 per cent of our GDP in 2002, it is too dependent on Singapore tourists. Out of the total 13.3 million tourists, 7.5 million or about 57 per cent were from Singapore while 76 per cent of the excursionists (day-trippers) were also from Singapore.

The moral of the story is that we must strenuously diversify our tourists and make our Singapore tourists far more welcome in the future! For instance, I cannot understand why it is beyond the ability of both sides to resolve the longstanding water problem with Singapore and remove this thorn in our flesh. Then tourism on both sides of the causeway will increase faster.

9. **Agricultural production and quality can be strengthened to reduce our imports and increase our exports.** We are fortunately blessed with abundant land and the expertise in primary commodities. But land administration and ownership are major constraints that must be overcome, otherwise we will fail to further tap our God-given rich natural resources.

10. **Finally, it will be unfortunate if we give priority to especially transport security when crime, corruption and personal safety are prime concerns as well.**

The Police and the Judiciary will need to be given more funds to enable them to reduce crime and to expedite the judicial process, for they provide one of the prerequisites of business confidence and socioeconomic well-being.

The new Stimulus Package cannot afford to be short-term in scope or even cosmetic. It has to be structural and even radical and show the way for a brave new future! *It will also help the future Prime Minister Abdullah Ahmad Badawi a great deal if the Stimulus Package is realistic, effective and implemented efficiently.* In the meantime, the threatening war against Iraq is dampening economic activities and blurring the international and domestic economic outlook. U.S. President Bush and U.K. Prime Minister Tony Blair appear to be determined to go for all out war against Iraq—because they claim that Iraq possesses WMDs.

U.S.-Iraq War and Economic Uncertainty

Both the Chief Weapons Inspector Hans Blix and the head of the U.N. Nuclear Watchdog agency Mohamed El Baradei reported to the Security Council on February 14, 2003, that they have found "no evidence of weapons of mass destruction in Iraq".

But the inspectors urged Iraq to increase their cooperation to carry on the search for the alleged WMDs.

Earlier, the U.S. Secretary of State Colin L. Powell's briefing of the Security Council did not convince even Pope John Paul II that the U.S. has "irrefutable evidence" of Iraq's possession of WMDs, according to the Pope's spokesman, Archbishop Renato Martino!

Nevertheless, the U.S. and the British leaders claim that the Iraqis have already committed "material breach" of the U.N. Resolution 1441 and that technically they can demand that the U.N. should endorse their war against Iraq!

Apparently, the previous Iraqi refusal to allow the U.S. to fly their U2 spy planes over Iraqi air space and the lack of complete and full Iraqi cooperation with the U.N. inspectors constitute some of the elements of this so-called "material breach" of the U.N. resolution.

But very few countries would support the U.S. and U.K. in going to war against Iraq—purely on these weak technical grounds of material breach!

Most countries state that as yet the U.N. inspectors have not produced any evidence of the presence of WMDs! Where are the "smoking guns" that were proudly announced by President Bush?

If they are hidden somewhere in the vast desert, why hasn't the U.S. superior air technology in surveillance and their intelligence service been able to find these WMDs?

The irony is that the U.S. which is doggedly aided and abetted by Britain, may actually be forced by their own folly to attack Iraq on one pretext or another, for fear of living up to the claim by Saddam Hussein's Baath Party newspaper, that President Bush is a "Super Idiot"!

The only way for the U.S. to win the War against Iraq—is not to start a War, but to help the U.N. inspectors to find the WMDs, if indeed there are any left!

If there is a war in Iraq or a even a quick massive U.S. missile strike on Iraq, the consequences can be damaging to the world economy! *The U.S. and its limited allies could win the battle, but lose the war in the longer term.*

Terrorism could be stepped up in the West and thus cause continuing international economic uncertainty and deflationary pressure. This possibility could undermine business and consumer confidence and restrain economic expansion!

It is therefore vital for some peaceful settlement of the U.S.-Iraq crisis. U.N. inspections need to continue their work and Iraqi cooperation has to increase so that the real truth as to whether Iraq has weapons of mass destruction or not will be found out. Then it will be easier for the whole world to deal with this highly dangerous world threat to peace and economic progress!

The U.S. war against Iraq will be very costly. But can the U.S. afford it, given its large budget deficits?

U.S. Budget 2004 for War

If war breaks out in Iraq, it will have to be swift, as the U.S. Budget and its economy cannot afford a long drawn war! The Budget 2004, starting from October 1, 2003, already plans for a deficit of US$307 billion as compared to US$304 billion for 2003! However, this estimate does not include the cost of a war against Iraq which is conservatively placed at about US$61 billion. This is the amount spent for the whole of the last Gulf War in 1991.

That is why the war has to be over fast. But if the war takes long to achieve its goals, then the U.S. Budget will be badly bashed up!

The Budget allocation US$36 billion for the newly established Department of Homeland Security will be grossly inadequate, if international terrorism steps up as a result of the U.S. attack on Iraq!

The Pentagon's allocation of US$380 billion will rise to US$400 billion in order to oversee the vast Nuclear Weapons Programme or its own WMDs, which, ironically, the U.S. does not want Iraq and North Korea and other countries to possess!

Because of the U.S. Presidential Elections in November 2004, the Bush administration has cut taxes by US$1.7 billion over the next 10 years. This will bring about an increase in U.S. debt.

White House Budget Director Mitch Daniels says it is "no great trick" to balance the Budget, except that it is necessary to battle terrorism and the weak U.S. economy. But how long can the U.S. economy thrive on borrowed money, because of the "safe haven" belief on the part of the world's investors?

What is to prevent these international investors from shifting their attention to Europe and Japan, where their investments would be much safer and less prone to attacks from terrorists who are bitter against the U.S. for its pro-Israel policies?

The U.S. will have to exercise more Budget discipline and come to terms with its distorted foreign policies. These ill-advised policies continue to support dictatorial governments and oppress poverty-stricken Third World countries through, *inter alia*, the IMF, the World Bank, and the WTO, in order to dominate the world!

The U.S. Budget is in a bad shape and could be self-destructive!

Greenspan Warns the U.S.!

The U.S. government has already been warned by Alan Greenspan of the weakening U.S. Budget position!

These are not mere speculations. The Federal Reserve Chief Alan Greenspan has told the U.S. Administration in his Congress hearings in February 2003 that, there is no need for a an economic stimulus, before any war against Iraq, particularly because of these mounting U.S. budget deficits.

He was critical of President Bush's proposals to Congress to cut taxes by US$695 billion, even before taking into account the vast sums already spent on amassing 150,000 U.S. troops and sophisticated military hardware in and around Iraq!

Greenspan's criticism was so severe that it prompted the head of the Americans for Tax Reform to say that "He [Greenspan] is getting too old"!

But that is the hazard that all central bank governors have to face all over the world—even in Malaysia! Central bank governors are supposed to be independent, but that is all well and good, as long as there is no serious difference of opinion with the government. However, in this case, Greenspan had the courage to come out openly to disagree with President Bush.

But I suppose he is old enough and does not need or want another term after serving devotedly for 15 long hard years!

Greenspan insisted that "running budget deficits does affect long-term interest rates. It does have a negative impact on the economy and has to be addressed"!

These warnings can as well apply to all countries indulging in continuing deficit financing, including Malaysia! However, in the meantime, our momentum to improve the quality of our national economic management must continue. But unfortunately, we seem to be losing some steam!

Examples of Malaysian Mismanagement

There seems to be some loss in the will to implement policies more effectively in Malaysia. A few unfortunate instances would give the impression that we are softening when we should be taking a tougher stand in managing the economy, given the growing international economic challenges.

1. **Water for Singapore.** Take the case of our problems with Singapore over our supply of drinking water to them. Why can't we work out a compromise between our requested price and their offer price for

our water. The differences are not so far apart. Its just the difference between 45 and 60 cents for 1,000 gallons of water supplied to Singapore by Johor.

Early settlement will take away the growing acrimony that is eating into the greater goodwill that the peoples of both countries have actually enjoyed for so long. The open public arguments on both sides can undermine business environment and confidence, which does nobody any good!

Surely the officials of both countries can devise an acceptable formula to settle the dispute even without having to go to arbitration. In the end its a question of dollars and cents and some goodwill and common sense!

There is no point in Singapore's Trade and Industry Minister George Yeo stating the obvious that "Our futures are intertwined", but not being able to pay us a fair price for the water they depend on for their survival. Instead Singapore is sticking to the letter of the law and not the spirit, in still paying the 1927 price of 1.5 Singapore cents per 1,000 gallons. It really makes no sense!

We do not want the people of both countries to express their dissatisfaction over this unsatisfactory state of affairs by protesting like the Cambodians and the Thais in regard to the claims on the Angkor Wat temples in February 2003.

2. **Hill Slopes.** Then there is the continuing problem of the blatant mismanagement and disregard of the laws and regulations pertaining to the development of hill slopes as evidenced in the recent case at Cameron Highlands.

How come the penalty is as low as RM500 for breaking the environmental laws and how is it that the authorities were so slack in enforcement? All these archaic laws of the country have to be updated as a matter of priority and the complacent officials should be severely disciplined. Greedy businessmen and errant officials cannot be allowed to betray the trust of the people, as they will react through the ballot box. We have to increase our will to manage more effectively.

3. **Non-*Bumiputera* Police.** The Police outreach programmes to attract more non-*Bumiputeras* into the Force is another problem of poor implementation. It is a sound policy to recruit more non-*Bumiputeras*, but has there been a thorough study to find out why non-*Bumiputeras* are reluctant to join the police force?

There is a public perception that non-*Bumiputeras* will not be given fair promotion prospects. What are being done to address these problems, before launching a recruitment campaign, that will probably turn out to be costly and fail?

4. **Train Derailment.** The recent train derailment at Gemas is another case of weak management. The Railways (KTM) and the Drainage and Irrigation Department (DID) blamed each other for the mishap. But this serious incident could have been avoided if each agency discharged its responsibility fully. Is there some deterioration in implementation and some dereliction of duty here too? There must be more pride in professionalism in our government system!

Kuala Lumpur International Airport

The same problem of effective management is seen at the Kuala Lumpur International Airport (KLIA), where the baggage system has broken down again—the second time since December 4, 2002! We have a world-class airport with third-rate baggage handling! Thanks to the intervention of the Minister of Transport Dr Ling Liong Sik, the baggage system is expected to be improved before the Non-Aligned Movement (NAM) Summit from February 20-25, 2003. But can't the system move effectively on its own steam?

The problem is mainly human, i.e., the lack of skills, training, supervision and poor work ethics. But what are the managers and workers doing to increase their productivity?

Productivity increases become more important as we simultaneously pursue our negotiations within the WTO, and seek to gradually become more internationally competitive. But with the looming war against Iraq, what is our priority for the forthcoming WTO negotiations?

Postpone the WTO Meetings

The Informal Ministerial Meeting of only 25 WTO members in Tokyo, Japan, on Valentine's Day, February 14, 2003, was not going to be an occasion for roses.

Minister of Trade and Industry Rafidah Aziz knew it well, for she has already declared, "I am not sure whether this meeting will contribute much to the Cancun meeting."

She was right. The formal 5th Ministerial meeting scheduled to be held in Cancun in September 2003 will also not amount to much.

How can there be progress in WTO negotiations when the major western WTO members continue to deny market access to the agricultural products of developing countries. At the same time the industrial countries continue to heavily subsidise their own agricultural products, against the vital interests of the developing countries?

In fact, this serious attitudinal problem of the Washington Consensus is one of the major causes of poverty together with the deep fear of U.S. imperialism and Western domination and oppression. All these provide a conducive climate for international terrorism to grow and prosper!

Hence the WTO negotiations should be put on hold!

Given the great uncertainty and the callousness shown by the U.S. and its allies against world opinion which is opposed to the impending U.S. war in Iraq, the WTO negotiations should proceed only if there is a genuine change in the mindset of the major Western countries.

The West has to decide to restructure the international trade and financial system to bring about socioeconomic justice between the rich and powerful and the poor and weak nations.

The WTO Ministerial meetings must be told and also understand that unless there is a radical change in their attitude and direction in the WTO, developing countries would have little enthusiasm to support WTO negotiations and agendas that benefit the rich much more than the poor.

The equity and fairness shown in the negotiations have a major bearing on the outlook for international terrorism.

Poverty, just like the severe consequences of the conflict between Israel and Palestine, will determine the direction and intensity of international terrorism in the future!

The U.S. National Strategy
for Combating Terrorism

Therefore it is commendable that finally, the U.S. has come out officially and clearly stated that the conflict between Israel and Palestine is a critical component to winning the war on terrorism and that "poverty is one of the underlying conditions that feeds extremism".

These major conclusions were expressed in the U.S. Report entitled "The National Strategy for Combating Terrorism" that was released in Washington on February 15, 2003!

The U.S. Administration had been in a dream world of denial in the past. It had previously taken the easy line that terrorism was caused by misguided Islamic fundamentalists.

The U.S. Administration had ignored the basic root causes of terrorism, which most of the world had recognised as—the illegal Israeli occupation of Palestine and world poverty. This is mainly caused by the exploitative and oppressive trade policies imposed by the Western-dominated international institutions like the IMF and the WTO!

The National Strategy Report outlines a "4D" Strategy to combat extremist groups, i.e., "to Defeat extremists, Deny them support or sanctuary, Diminish the underlying causes and to Defend the U.S."!

Solutions to Combat Terrorism

To achieve these laudable goals, however, the U.S. and its allies have to fundamentally change their basic political and economic policies!

The U.S. cannot win the war against terrorism through war but by winning the hearts and minds of the oppressed peoples of the world. This can be achieved by:

Firstly, the U.S. has to be neutral in its handling of the Israel-Palestinian conflict and not employ double standards by backing Israel against Palestine!

Secondly, the U.S. and the rich and powerful industrial countries have to adopt and implement international economic and financial policies that equally benefit the developing countries. The U.S. and its allies have to abandon perpetuating old and outdated international trade and financial policies, that promote the domination and hegemony of the developed West over the undeveloped South. *Only then can the U.S. strategy to combat terrorism really succeed.*

The so-called seven state sponsors of terrorism, that the U.S. has identified as Iran, Iraq, North Korea, Cuba, Libya, Syria and Sudan, will not be encouraged to join the self-acclaimed "U.S.-led coalition against extreme violence", as long as Israel itself and some other Western allies are not acknowledged as state sponsors of terrorism.

Indeed the U.S. strategy to combat terrorism will fail, even if the U.S. commits the extreme blunder of attacking Iraq, if urgent action is not taken to address the real causes of terrorism, i.e., the Palestine tragedy and poverty in developing countries. If the U.S. and its allies have the sincere will, there will be many ways to win the war against extremism and terrorism. The world prays and waits for peace to prevail.

The Non-Aligned Movement Summit held in Kuala Lumpur in February 2003, has also added its support to the "No War in Iraq" worldwide campaign. But it is uncertain at this stage whether the U.S. and the British are going to rush into war with Iraq, where angel's fear to tread!

The NAM Summit, February 20-25, 2003
The Non-Aligned Movement's 13th Summit issued a balanced and substantive Kuala Lumpur Declaration on February 25, 2003. The Conference was attended by 114 countries represented by Heads of Government and Heads of State.

The theme of the Summit was "Continuing the Revitalisation of NAM". It stressed the need for NAM to take a more proactive role in international issues. It also agreed to strengthen the United Nations and multilateralism, as opposed to the present U.S.-led tendency to adopt unilateralism, as in the case with the probable U.S.-led invasion of Iraq!

In the six-page Declaration, the NAM leaders, *inter alia*, urged members to strengthen national capacities, enhance South-South cooperation, and to promote more dynamic and cooperative relationships with the developed industrial countries.

Most importantly, however, was the decision to "strengthen the role of the chair, as spokesman for NAM, through the establishment of appropriate mechanisms as part of the necessary backup system."

This decision could be very significant, considering that Malaysia will be the Chair for NAM for the next three years!

Dr Mahathir said he will be the Chairman for only the next eight months as he had publicly announced that he will retire in October 2003. Then the present Deputy Prime Minister Abdullah Ahmad Badawi will take over as chairman of NAM!

By then it is hoped that Dr Mahathir would have firmly established some mechanism or a secretariat that would be able to effectively consult all the 116 NAM members and coordinate their action to play a more proactive role in international political and economic relations. This is essential especially at world fora like the U.N., the WTO and the IMF!

Already I can see that there are some ideas floating around that the new NAM style could be based on Malaysia's impressive success in organising the Langkawi International Dialogues (LID).

The Southern Africa International Dialogues have also adopted the Malaysian Langkawi International Dialogue model which has proven successful.

The Langkawi International Dialogues, that have been designed by the Malaysian Industry-Government Group for High Technology (MIGHT) together with the Commonwealth Partnership for Technology Management (CPTM) in London, have promoted its philosophy of Smart Partnership and Prosper-Thy-Neighbour economic and social policies that aim for win-win outcomes in international relations.

NAM Goals

This philosophy could well be used for the promotion of stronger NAM ties in the achievement of the original NAM goals which are as follows:

1. Decolonisation;
2. Disarmament;
3. Development; and
4. I would add a new goal: Defence against political and economic terrorism, both individual and state terrorism!

The five original NAM Principles are also still relevant, i.e.:

1. Territorial integrity;
2. Non-aggression;
3. Non-interference;
4. Equality and mutual benefit; and
5. Peaceful co-existence.

All these goals and principles are remarkably still very pertinent (or more so now) in a unipolar world economic order where the balance of forces under the old bipolar order has gone, with the collapse of the Soviet Union.

The U.S. and some Western powers, particularly the U.K. that is the chief U.S. ally, will now reinforce their aim to dominate the world, by the use of their superior military might.

They would use it unscrupulously, as in the case against Iraq, to achieve their dubious aims as the new imperial powers to dominate the Third World!

They would dominate world trade and finance through the control of the international agencies like the WTO, the World Bank and the IMF and through the strategy of globalisation.

The Third World, without military might and without any other superpower for support, will be left to the mercy of the superpowers and their Western allies. They must therefore unite to resist this growing domination of the South by the North!

NAM countries in the South could therefore provide this defense if the new chairmanship under Malaysia takes the lead to initiate new measures and institutional changes to make NAM much more effective than in the past.

Malaysia has a great challenge as well as a great historical opportunity, to serve the Third World. I hope we will be able to rise to the these challenges to overcome these international threats of world domination by the rich and powerful over the poor and weak nations of the world! But Malaysia has to overcome its own problems first.

CHAPTER 5

ECONOMIC PROBLEMS
AND RESPONSES

ALTHOUGH Malaysia's economic fundamentals are basically healthy, the economy is thought to be unlikely to achieve the Budget target of 4.0 per cent for 2002. Neither is there real confidence in the outlook for 2003 at the present time, but the international economic outlook could improve.

However, Bank Negara Malaysia reported its encouraging assessment that the Malaysian economy registered a higher than expected 5.6 per cent growth in the third quarter of 2002. If the fourth quarter performs as well, then the Budget target of 4.0 per cent growth may just be achieved, given the high government consumption spending which grew by 21.7 per cent in the third quarter!

It is more important for the private sector, especially domestic and foreign investment, to contribute to economic growth for long-term sustainability.

The World Bank, in its mid-year review of East Asia and the Pacific entitled "Making Progress in Uncertain Times", estimated economic growth for Malaysia in 2002 at 3.5 per cent. It highlighted Malaysia's sound economic indicators, like the comfortable foreign exchange levels, low inflation at only 1.9 per cent, and the healthy balance of trade.

However, the World Bank cautioned about the challenges ahead such as the rising public debt and the slowdown in private investment and foreign direct investments (FDIs).

To these concerns must be added the continuing budget deficits and the balance of payments strains, especially on the current account. The fast growth in public consumption cannot be as productive as economic investment. *Most importantly, national productivity and integrity have to be raised.*

These warnings should be taken seriously and policy measures should be developed now to reduce their negative impact on the economy.

Prime Minister-designate Abdullah Ahmad Badawi will have to overcome these economic problems that he will have to deal with later on when he assumes power.

In the U.S., the largest economy in the world, the economic growth for the third quarter of 2002, was only 3.1 per cent at an annual rate.

The U.S. Commerce Department Report underlined the "the uneven pace of recovery". Already in October, the sale of cars and trucks have slowed down, despite free financing!

The U.S. Commerce Secretary Don Evans in commenting on his department's report, said that there were a number of influences that were "depressing consumer spirits", including anxiety of corporate accounting scandals and a possible war with Iraq!

In fact, U.S. consumer confidence has slumped to a nine-year low in October 2002! Furthermore, a U.S. Labour Department report indicated that the rate of increase for wages, salaries and benefits, has declined to 0.8 per cent in the 3rd quarter, compared to the previous quarter of 1.0 per cent!

Indeed the U.S. economy could well be slipping into a deflationary trap, like the Japanese economy!

This trap is caused by a lack of confidence to spend, by both private consumers and investors, when there are growing U.S. threats of war in Iraq. It is also a matter of insecurity as well, since there is so much of a "siege mentality", resulting from the fear of terrorism in the American homeland itself!

The answer is for the U.S. to spend itself out of this economic malaise. But how much can government spend and for what viable purpose and how soon can this take place.

In the meantime the U.S. economy needs a strong shot in the arm, to lift the economy and propel it forward, before it slows down further and slides more steeply.

Some analysts believe that this vital boost to the U.S. economy could be provided by a full-blown war in Iraq and even in North Korea!

That would be a terrible price to pay for the world at large, although the vast U.S. defense industry and big businesses like U.S. oil and steel would benefit greatly from war!

The U.S. Republican Party which is financed by Big Business is undoubtedly forcing President George Bush to go to war, but hopefully the rest of the world and even America's staunch ally Britain, will be able to stop U.S. belligerence and this insanity of going to war in Iraq?

The U.S. reason to attack Iraq is now unnecessary since Iraq has accepted the U.N. Resolution 1441, which allows U.N. inspectors, to "go anywhere at any time, to search for weapons of mass destruction". Non-compliance will lead to "serious consequences"!

The U.S. has repeatedly asserted that it is prepared to go to war if there is non-compliance with the U.N. resolution. This stance has lead Iraq and most developing countries to believe that the U.S. wants to go to war mainly to control the oil supplies in Iraq and to dominate the Middle East together with Israel, that has greatly influenced U.S. foreign policies.

The U.N. Resolution now expects Iraq to provide a complete declaration of all aspects of its biological, chemical and nuclear weapons programmes by December 8, 2002. Then we will see what happens! *An incomplete list would give the U.S. a good excuse to attack Iraq. However, if the list is complete, the U.S. can still find some fault, to lead a Western coalition attack against Iraq!*

In the meantime Iraq has repeatedly claimed that it has and will not have weapons of mass destruction (WMDs).

Hans Blix, the chief U.N. weapons inspector, and Mohamed El Baradei, the head of the International Atomic Energy Agency (IAEA), will have the grave responsibilities, of ascertaining whether the U.S. and its allies are truthful in claiming that Iraq indeed has WMDs!

If after all these accusations and denials, the U.N. inspectors declare that Iraq has no WMDs, as Iraq had always claimed—'then the world would see the greatest hoax perpetuated by the U.S. and the U.K.!

The world therefore waits with anticipation for the real truth behind these claims and counter claims on whether Iraq actually possesses WMDs!

The truth shall make the global economy free to move forward, either after a quick war in Iraq or even hopefully without a war!

The best scenario would be for no war. Then the U.S. and the rest of the world could better concentrate on the real root causes for the threat of international terrorism, which is political oppression and economic exploitation that causes poverty.

Then, international confidence will bounce back, and give our Malaysian economy the necessary fillip to leap forward on a sustained basis.

However, nearer home, there are also worrying signs that Malaysia's economic progress is being hampered by uncertainty and the dampening consumer and investment climate. Thus the continuing spat between Singapore and Malaysia over the longstanding grievances over Singapore's payment of a fair price for Malaysia's supply of drinking water to Singapore, does not provide confidence for stronger economic cooperation between our two countries for higher economic growth.

The War in Iraq

The Malaysian economy is facing economic challenges on two fronts. These challenges must be overcome so that the transfer of power, from Dr Mahathir Mohamad to Dato' Seri Abdullah Ahmad Badawi, who is to become Prime Minister in November 2003, will be as smooth as possible.

The first challenge is the adverse economic impact of the war in Iraq and secondly the battle in the WTO trade negotiations!

On the domestic economic front, Bank Negara's well-presented Annual Report for 2002 estimated that the Malaysian economy will expand by 4.5 per cent in 2003 and by 6 to 7 per cent in 2004. This confidence is based on the assumptions that "there is no stress or strain in the economy that can make the economy worse" and that "there will likely be a shift in the driving force of the economy this year [in 2003], from the government to the private sector."

However, the budget deficits for the last five years could be a major cause for strain. Furthermore, the private sector may not be able to provide the driving force that is expected to compensate for the slower public spending of 3.6 per cent compared to the 8.6 per cent expenditure incurred in 2002.

Private consumption is estimated by Bank Negara to increase by 6.9 per cent in 2003 compared to only 4.2 per cent in 2002. Private investment is expected to expand by 8.1 per cent in 2003 compared to the dismal negative 6.1 per cent in 2002. These are ambitious projections which would be difficult to realise, unless the war ends abruptly and we really liberalise the economy soon. But these assumptions are unlikely to materialise, judging from the present international and domestic developments!

But Bank Negara was right in indicating that, if the war in Iraq extends beyond one month, then Malaysia's strong fundamentals and the growth estimates could be adversely affected.

Then the present flexibility to respond to international changes will become more limited and the risks would increase!

Therefore the actual duration of the invasion against Iraq by the U.S.-led coalition that started on March 18, 2003, will be crucial for the world and Malaysian economy. The war in Iraq will have adverse effects on the world as well as the Malaysian economy.

Indeed, the U.S.-led war of the small "Coalition of the Willing" of only the U.S. and two other large countries—Britain and Spain, and about 35 smaller, less significant countries that are indebted, could cause world deflation.

Therefore the assertion by Bank Negara that policies in 2003 will create "an enabling environment" will be very assuring to businessmen and investors—if there is sufficient liberalisation!

But we will have to wait for the government's Stimulus Package that was expected later in March 2003 (but which has not yet appeared by the end of March 2003), to find out how much more encouraging the package would be for the private sector.

But now the announcement of the package has been postponed to April 2003. However, regardless of the Iraq War, we would have needed this Stimulus Package much earlier!

It was previously hoped that the war against Iraq, if it had to happen, would be sharp and swift, lasting perhaps a few days. *But the war has gone on for some time now—and the Battle for Baghdad is still not over as at the beginning of April 2003, about two weeks after the war started!*

How long more will it take the U.S. military might, together with the strong support of its faithful followers like the British, the Spanish and even the Australians—to subdue the poor economy and inferior military strength of Iraq?

Using a sledgehammer to kill a fly has not helped much to win the war—and especially the war for the hearts and minds of the Iraqis and the Arabs and indeed most of mankind!

The U.S.-led war in Iraq will certainly not help to fight the war against international terrorism, which is really a war against imperialism and political oppression and economic injustice, that has been inflicted by the rich and the powerful, against the poor and the weak countries!

In fact, international terrorism could now gain more sympathy and support and further strain the world economic growth.

A longer war can be even more debilitating not only to the countries engaged in the war but all the other peaceful countries! Economic uncertainty and instability could increase rather than decline after the war against Iraq! The war therefore must cease soon.

Already the world economy is suffering with the sharp fall in tourism, the near collapse of some international airlines, and the depressing effect on consumption and domestic investment and on FDIs.

Add to these depressing economic trends is the staggering increase in non-productive spending on the war itself, on defense, and on anti-terrorist activities, especially in the war-like countries—which are also among the most developed and richest industrial countries.

We cannot isolate ourselves from the war in Iraq, as suggested by *New Straits Times* columnist P.Y. Chin on March 23, 2003, as Malaysia is the 18th largest world trading partner.

But we can compensate for its ill effects on our economy, by stepping up the pace of change in policies and processes, to stimulate and sustain stronger economic growth and better income distribution or distributive justice.

WTO Negotiation Challenges

The second challenge is the battle going on at the WTO negotiations.

While the war in Iraq is raging, the WTO is striving to advance its trade negotiations! But which warring country can give high priority to international trade, agriculture and investment issues, when they are locked in battle!

Hence, the WTO deadline for further commitments in market access and tariff cuts that were fixed for February 17, 2003, has been overtaken by events!

Similarly, the negotiations to eliminate all non-agricultural tariffs by all WTO members by 2015 is also now moving even more slowly.

For the service sector, the industrial countries had aimed to table their offers for more liberalisation by March 31, 2003—but that date has passed.

Nevertheless, this proposal will mean nothing as the developing countries are not ready to table their offers for opening up the service sector. The poor countries fear the rich and powerful industrial countries will use their overwhelming comparative advantages and technology to unfairly compete with the developing countries—as in the case of their military superiority in the war in Iraq!

The perception of the developing countries is one of suspicion and that the rich countries are out to get them as they did in the colonial era!

Hence there appears to be delay if not a stalemate in the WTO, particularly with the U.S. giving more priority to fighting in Iraq and against its many adversaries, including international terrorists, all over the world!

But this may be a temporary respite in the WTO negotiations in Geneva, Switzerland. *As soon as the war in Iraq is over, the U.S. will step up its efforts to strengthen its economic power by squeezing more concessions from the rest of the world and especially the Third World, through the WTO negotiations.*

Thus Malaysia and other developing countries must work closely to resist undue pressure to undermine the principle of "progressive liberalisation" which is provided by the WTO rules.

More importantly, Malaysia has to increase its own internal consultations, coordination and consensus, on an agreed timetable for the liberalisation of all sectors of the economy, according to our capacity and national interests.

This goal cannot be achieved by any one Ministry or industry alone. It has to be a national project where all sectors of business are fully committed and actively engaged.

But this message does not seem to have permeated widely enough throughout the public and even the private sector, which has the most to lose, from little realisation of and even less participation in, the laborious process of WTO negotiations.

Fortunately, the National Economic Action Council Secretariat led by Dato' Mustapa Mohamed, has been holding a series of intensive meetings, with the private sector and government agencies, to step up the awareness and coordination among a whole range of ministries, departments and business leaders, to back up and enhance our WTO negotiating capacity.

Even our Acting Prime Minister Abdullah Ahmad Badawi had to advise Malaysian law firms, in his address to the Malaysian Bar on March 22, 2003, "to act as reliable support for Malaysian business ventures abroad, as well as to target potential businesses intending to invest in Malaysia"!

But how can Malaysian lawyers be able to effectively provide "reliable support", when they practice relatively "closed-door" practices in protecting themselves against foreign lawyers, by making it difficult for them to practice here in Malaysia.

Thus the foreign Bar Councils will similarly prevent our lawyers from practicing in their own countries.

If this protectionist trend continues, among lawyers, and most other professional groups like the architects, doctors and engineers, then Malaysian professionals will not gain the competitive edge that is so essential to provide professional services of world-class standards!

Hopefully, there will be greater priority given by all sectors of our economy to face the war in trade and investment negotiations in the WTO.

This can only be achieved if the government sends clear signals to the business community and particularly to the professional groups, that protective policies will be phased out, to force them to be more competitive internationally as soon as possible!

However, if the government itself is reluctant to do so for populous and political reasons or mere inertia, then we should not expect to increase national productivity and to become internationally competitive enough. Then we must accept to decline in our economic performance and not expect much more progress!

To win this WTO war, we therefore must resolutely prepare urgently (which is not clearly evident yet), for more competition internally, to be able compete better internationally.

Otherwise, our economy will lose out on both fronts—not only from the war in Iraq but also the war in the WTO, in our struggle to maintain and improve the status quo—and particularly, to prevent more economic slide in the longer term!

Thus the government has to change its own mindset, just as we should all alter our mindsets for Malaysia to continue to prosper and succeed!

This transformation of the national mindset must take place at a faster pace. There are just over five months left before November 2003, when Abdullah Ahmad Badawi takes over as Prime Minister.

Hopefully the Malaysian mindset will change quickly, so that the new Prime Minister will be able to take off with a flying start, rather than have to experience a slow start!

The Malaysian Malaise that Abdullah described so courageously at his recent Oxford-Cambridge breakfast talk should not persist to thwart the new Prime Minister's determination to carry the torch from Dr Mahathir and to run with it towards industrial status by 2020!

Malaysia has therefore to adopt more dynamic policies not only to prepare and to get much more from the WTO negotiations, but to deepen our bilateral trade and investment with other countries, especially with those that we have comparative advantages.

Such a country is Thailand—our close adjoining neighbour, with whom we have had longstanding fraternal relations.

Malaysia-Thai Cooperation

It is encouraging to note that Minister of International Trade and Industry Rafidah Aziz has led a successful trade and investment mission to Thailand.

In her speech to the seminar on "Malaysia-Thailand Business Opportunities", in Bangkok on February 28, 2003, she expressed strong confidence in promoting greater trade, investment and tourism between the two countries. Indeed, it is a pity that so far our trade and investment has not been as strong as it should be. In 2002, Thailand was only the 7th largest trading partner for Malaysia. Total trade between the two countries amounted to the small volume of US$7.1 billion. Approved investments by 65 Malaysian companies in Thailand totalled only US$39 million in Thailand in 2002. For the last 5 years Malaysian investment in Thailand amounted to only US$1.0 billion!

Worse still, for the same period 1998-2003, approved Thai investments in Malaysian Manufacturing amounted to a disappointing US$25 million!

73

The Thais apparently find it less attractive to invest in Malaysia while our Malaysian investors find it more rewarding to invest in Thailand.

Why is this so? MITI could investigate the reasons in order to enhance trade and investment between the two countries.

There could be a growing trend for Malaysian businessmen to invest in other AFTA countries and a declining interest for foreign as well as Malaysian businessmen to invest in Malaysia.

Could this be happening perhaps because of our more restrictive investment policies or our relatively less friendly business environment?

We have to seek new ways and means to increase bilateral trade, investment and tourism, between two friendly neighbours such as Thailand, especially since Singapore has a hostile attitude in their bilateral relations with us.

For instance, in February 2003, Singapore's Foreign Affairs Minister S. Jeyakumar told the BBC that the Singapore approach is to resolve the outstanding water issue, in accordance with international law and in compliance with existing agreements.

Thus it appears that they are not interested in positive and proactive negotiations to break the present unnecessary stalemate.

If that is Singapore's attitude then it becomes more important for Malaysia to look more closely at our more sincere Asean neighbours, to strengthen and expand our regional trade and investment. We should gradually move away from those trading partners who undermine our economy and the Asean fraternal spirit.

With AFTA, we have to review and revise all our trade and investment relations with greater priority and pragmatism.

However, we have to change our traditional mindset which has kept us tied to old trading partners from colonial times, mainly because we have not been adventurous and innovative enough, to remove our "ugly values".

The Ugly Malaysian

The *Sunday Star* editorial on March 9 rightly commented that the Cabinet Committee on Competition headed by the Acting Prime Minister Abdullah Ahmad Badawi, "is what the country needs at its current stage of development".

Indeed, the Competitive Committee should have been established much earlier. But *The Star* editorial added "it is not too late, if we can still recognise the faults in ourselves"!

On his first morning as Acting Prime Minister, Abdullah Ahmad Badawi, on March 6, 2003, gave an auspicious speech to the Oxbridge Society and stated boldly that "most importantly, we need to think differently," in order to achieve the goals of Vision 2020! He added that "we must compete for it and work hard to achieve it". He signalled that Dr Mahathir's Vision 2020 had become his vision but also indicated that our road map had yet to be clear cut.

As chairman of the Competitive Committee he would therefore seek to increase meritocracy among individual Malaysians, urge greater competition among corporations, whereas the government should become more of a facilitator.

Abdullah Ahmad Badawi is spot on, but we need to know how we got to this stage of developing a Malaysian Malaise. We have First World infrastructure but a Third World mentality in the maintenance and administration of these public facilities?

The answer is that some of our leaders had inadvertently allowed our human quality and performance to deteriorate, by the wrong implementation of our "redistributive policies"!

In other words, the manner of carrying out the affirmative-action policies and basically the NEP has gravely fallen short of the expectations of its founding fathers!

Hence we have a large number of "Ugly Malaysians" in our midst. They are people who are corrupt, disrespectful of public property and property rights, and who believe that "the government owes them a living"!

Thus we have to change from being complacent to becoming much more competitive in a fast globalising world, otherwise we will be "*gobble-ised*" and suffer national decline.

I would agree with Acting Prime Minister Abdullah Ahmad Badawi that the key to the change of our mindsets must therefore be to base our policies on more competition, merit and excellence.

As he pointed out, only those with "genuine need and those who have value-added potential" should now be given assistance to attain redistributive justice.

But this noble goal was in fact intrinsic in the NEP, which postulated the eradication of poverty regardless of race and the removal of the identification of race with occupation!

However, many political leaders and civil servants have somehow, over the last 25 years, distorted the true spirit of the NEP in the very implementation of the policy!

The distortion of the implementation of the NEP is the main reason for the growing polarisation in our country. All races have felt a sense of alienation as the few have benefited from "unproductive economic rents" and the majority of all races have not adequately benefited from "know how" but "know who" under the NEP.

As the *New Straits Times* editorial stated on March 7, "History has determined therefore that Abdullah will take over the controls of a nation and economy back on track and firing well on all four cylinders."

But the irony is that the Malaysian economy need not have veered so much off track in its values and mindsets, if not for selfish gains and greed.

Fortunately, however, it is not too difficult to change the course of the Malaysian economy and to change the mindset and to remove the Malaysian Malaise.

What we need is to sharpen our focus on greater competition and meritocracy to "move up the value chain", even if it means to "push ourselves in ways we have never imagined"!

Thus Acting Prime Minister Abdullah Ahmad Badawi, having identified the causes and effects of the Malaysian Malaise and the elements of the Ugly Malaysian, will hopefully present a new plan to the nation, on how to realise Vision 2020.

The Civil Service: Challenges Ahead

It was therefore refreshing to read former *New Straits Times* Editor-in-Chief Tan Sri Abdullah Ahmad's candid comments to the senior civil servants in Perak on the subject, "Shaping an Excellent Civil Service: Challenges Ahead", on March 7, 2003.

Indeed as Abdullah Ahmad pointed out, "What will become of us if the civil service of the future is manned by third-raters?"

We have not fallen to that dismal state as yet, but we may be moving fast in that direction, especially at the lower levels of the public service.

If and when the quality of the public service declines to "third rate", we can assume quite safely that we will lose sight of Vision 2020!

But it will be disastrous for us all if this should unfortunately come to pass. Hence we must find the causes of this malaise in the public service and overcome the weaknesses urgently.

The Declining Public Service

The causes of the declining public service are manifold.

Firstly, the public service is no longer attracting the best brains in the country, whether in the Administrative and Diplomatic Services or in the Professional and Technical and Clerical Services. This is because of the less attractive salaries and terms and conditions of service, compared to the much more competitive and financially rewarding private sector, primarily at the most senior levels.

So let's pay more for better quality. It would be cheaper than paying more for mere quantity and poor productivity.

Secondly, the public service has become almost mono-ethnic and so too much in-breeding has set in. The mindset has consequently become one of perpetuating mediocrity rather than meritocracy.

The solution would be to ensure that the public service becomes more multiracial to reflect the ethnic composition of our society. There appears to be a built-in bias in the mindset of many government recruiters not to employ more non-*Bumiputeras*.

Perhaps some kind of balanced quota should be introduced to recruit the many qualified non-*Bumiputeras* who have got fed up with being rejected in their applications for government jobs.

Thirdly, the government must make it clear that they want a highly qualified and competent public service, at least at the highest levels of management. Promotions should be strictly based on performance, on "*know how*" rather than "*know who*"!

Fourthly, the government must give greater priority to the professional independence of the public service, particularly at the most senior levels, without undue political interference.

Finally, discipline in the public service must be enforced with a stronger will and made easier to administer, without protection and preference. Thus the existing cumbersome disciplinary and dismissal procedures should be revised and modernised.

However, if we do not take resolute measures now to improve the quality and performance of our one million public servants, we the taxpayers will have good reason to worry about "what will become of us", and indeed our beloved country.

I hope our Acting Prime Minister, who was an outstanding civil servant, will soon be able to change the mindset and malaise in the public service and remove the image of the ugly public servant.

All Malaysians will surely support Abdullah Ahmad Badawi wholeheartedly in his mission to improve the public service and our national mindsets so as to achieve the goals of Vision 2020.

The challenges facing the Malaysian economy can be overcome only if there is greater national unity and a far greater commitment to quality and international competitiveness.

Abdullah Ahmad Badawi has only about five months before he assumes the office of Prime Minister of Malaysia, after Dr Mahathir Mohamad retires in October 2003.

The next six months are therefore crucial for Abdullah to prepare for the transfer of power and for succession. This transition period must be devoted to building Malaysia's socioeconomic and political resilience, international competitiveness and national unity, so that the transfer of power to Abdullah will be as smooth and as sustainable as possible in the longer term.

The expected NEAC Stimulus Package could contribute substantially to sustainable economic resilience and to strengthening the impending transfer of power to the new Prime Minister Abdullah Ahmad Badawi on November 1, 2003. But the war in Iraq is worrying!

"The War in Iraq will Cause Economic Disaster," says Dr Mahathir

Prime Minister Dr Mahathir Mohamad came out strongly to state that the war in Iraq is already causing economic disaster to the whole world!

In one of his most severe criticisms of the U.S.-led coalition's invasion against Iraq, he told the Arab TV network Al-Jazeera on April 5, at the Prime Minister's residence in Putrajaya (while on two months' leave from office), that *"the only way for America to stop [the war in Iraq] is for their own people to throw out their government"*! He also said that the "U.N. at the moment, is totally ineffective, if not useless, because it has failed to uphold the law and it has no means to uphold the law"!

Dr Mahathir's other strong criticism is listed below for the record, as it has an important bearing on our foreign economic policy and our future relations with the U.S. and the U.K. Dr Mahathir's main criticisms are outlined as follows:

1. **Objectives of the War.** Dr Mahathir believed that the objectives of the war against Iraq is to ensure that Iraq becomes totally incapable of defending itself at any time in the future. And, of course, there will be the oil they can get from Iraq. But mainly, I think it is to fulfill Israel's objectives, which is to eliminate any threat to Israel in the Middle East.

   I believe that Dr Mahathir is right in stating that the real motives of the Anglo-American attack on Iraq is to impose their will and hegemony over the Middle East and to protect their protegeé Israel. After all, the Jews have disproportionate influence both in New York and in London through their extensive financial control of the world's money markets.

   The other major reason is the vast oil reserves that are available in Iraq which is the world's second largest oil producer! The U.S. itself is a wasteful consumer of petroleum and needs access to oil reserves for its survival and particularly for its world dominance. *Hence the U.S. is anxious to control these vast oil reserves.*

2. **Media Bias.** According to Dr Mahathir: "They [the U.S. and the U.K.] have been telling a lot of lies throughout the war. Viewers are being denied the truth about the war." He quoted the case of former CNBC correspondent Peter Arnett who has been removed because he said something they did not like.

   My view is that it is understandable for the Western Press to be biased as it is owned by Big Business in the U.S. and the west. The mass media has therefore to be slanted in favour of the Western owners, who pay their salaries and who call the shots. It is the old story of the piper playing the tune, and the rats following.

3. **Reconciling trade with the U.S.'s and Malaysia's criticism against the U.S. on the war in Iraq.** The Prime Minister said that he would like to maintain Malaysia's share of the market with the U.S. However, "if we are subject of an attack, all the good relations and trade will come to nothing"!

I think we have to be more careful here as the U.S. can become vindictive as they have said that those who are not with them are against them!

Malaysia is perhaps the only developing country that has been so critical of the U.S. Even China which is a permanent member of the Security Council has been relatively mild in its criticism of the U.S. on its attack against Iraq! India and other big developing countries are similarly constrained in their criticism of the U.S. and the U.K.

Malaysia takes a tough stand on principle and believes that it is one of the few countries that can do so, because it does not receive any aid from the Anglo-Americans. Neither is Malaysia seriously indebted to the U.S. and the U.K. All its debts, it pays on time and even often repays them ahead of time. Thus the U.S. and the U.K. are in no position to "put the screws" on Malaysia!

Nevertheless, the U.S. could bide its time and when the war in Iraq is over, the U.S. could make it difficult for Malaysia in trade and investment!

4. **Rebuilding Iraq.** *Dr Mahathir mentioned that "The economy is going to be so bad that talking about rebuilding Iraq would be hypocrisy, because nobody can rebuild Iraq, just like they cannot rebuild Afghanistan"!*

Of course, the rebuilding of Afghanistan has been more difficult because there has been no immediate revenue from oil, to help finance the American and British contractors. But it is very different in the case of Iraq, where oil revenues are plenty to finance coalition contractors.

This is where the spoils of war will have to be shared. But the U.S. and the U.K. will want to take the lion's share. They say they will give the U.N. a "vital role" in the management of postwar Iraq.

But it is highly improbable that they will give any other U.N. member a share of the spoils, especially all those who had the courage to oppose the unilateral invasion by the U.S. and U.K. on Iraq, against the will of the U.N. Security Council and about a hundred U.N. members!

5. **Recolonisation.** The Prime Minister explained that "Unfortunately, today's colonisation takes many forms, among which is economic hegemony." He added that NAM countries "face a new phenomenon in which there is no respect for morality, no respect for international law and in the face of this, NAM will find difficulties in trying to maintain the independence of member countries."

In the aftermath of the invasion of Iraq, it is likely that the U.S. and U.K. will trumpet their success in having "liberated Iraq" from the tyranny of Saddam Hussein, but they will down play their real aim which was to exploit the oil wealth in Iraq, rather than liberate the Iraqis"!

But what about the original aim of discovering and destroying the so-called WMDs? Has that basic aim been forgotten so soon.

Was it therefore an unprecedented ruse in the first place? Was it a dishonest excuse to frighten some countries into agreeing to support the U.S.-U.K. invasion, in order to 'dominate' the Middle East? *Was it the Zionist movement, the Washington Consensus or an Anglo-American conspiracy to "rule the world"? The truth will surface one day. Only time will tell!*

Time will vindicate the real villains. If the U.S. and the U.K. leave Iraq and allow the Iraqis to govern Iraq themselves immediately after the war ceases, then I would give the Americans the "benefit of the doubt". This is what the U.S. Ambassador to Malaysia Marie Huhtala has asked for in her interview with Bernama, the Malaysian National News Agency, on April 6, 2003.

She added that "the U.S. is doing what it believes is the right thing". But how could it be the "right thing" when only 49 out of the 190-odd U.N. members supported the U.S. in some way or other. Only Britain in a big way and, Australia, and Poland to a much smaller extent, provided troops to invade Iraq!

Ambassador Huhtala mentioned that Iraq had failed to disarm its weapons of mass destruction (WMDs) despite 17 U.N. Resolutions. But she did not mention that so far no WMDs have been found in Iraq—by the U.N. inspectors nor the occupying U.S.-U.K. military in Iraq!

Furthermore, while Iraq has defied 17 U.N. Resolutions, the Israelis have defied about 30 U.N. Resolutions over a longer period of time—and yet the U.S. government continues to provide the Israelis with military aid and strong support in suppressing the Palestinians!

These are double standards practised by the U.S. and its faithful allies like the U.K. and Australia that the U.S. Ambassador has chosen to ignore! This is the kind of U.S. diplomacy that is dishonest and which causes resentment and a lack of credibility in the integrity of the U.S. and its allies!

If the U.S. and the U.K. continue to stay behind in Iraq and hand out most of the oil concessions and reconstruction contracts to U.S. and U.K. businessmen and their oil companies, then we will see the true colours of the Americans and the British.

We will soon be able to ascertain the truth of their real motives in wanting to "liberate" the Iraqi people from the tyranny of Saddam Hussein, or whether they wanted to "liberate" Iraq from its rich oil reserves and supplies!

In the end, the truth shall set us free. Similarly, we have to find out our own truth in Malaysia over the longer-term viability of the national car!

Can the National Car Survive?

The 500,000th unit of Malaysia's own Perodua Kancil car was proudly rolled out of the assembly-line at a ceremony officiated by Minister of Trade and Industry Dato' Seri Rafidah Aziz on April 11, 2003. But for how long more can it roll out and survive, after the car tariffs are reduced to only 5 per cent for cars imported from Asean countries under AFTA rules, on the extended date of January 1, 2005?

As the Minister of Trade and Industry rightly pointed out, "At this point in time, it is immaterial how much and when the tariff structure will be undertaken"!

Apparently some auto industry players are waiting for the government to announce tax measures "to compensate for the revenue loss" resulting from the drastic lowering of the current high protective tariffs.

But when was the promise to compensate the auto industry for loss of revenue given? And why should the auto industry be compensated and subsidised for so long by the taxpayers?

It is only proper, as the Minister of Trade and Industry frankly pointed out, that the auto industry has to operate with a mindset that is competitive and must benchmark itself against its regional and international rivals.

But can the auto industry make it? The senior managing director of Daihatsu Motor Co., Kentaro Shimizu, stated that Perodua, the national carmaker that produces the Kanchil, the Kembara, the Kenari and the Kelisa, is trying very hard to achieve its corporate slogan—"Towards 333". This entails attaining the goals of:

1. Reducing costs by 30 per cent by 2006;
2. Reaching the top three in the Customer Satisfaction Index by 2004; and
3. Carving out 30 per cent of the AFTA market share.

These are worthy and ambitious targets. But what assurance have we got that these targets are realistic and attainable? If these targets are not achieved on time, will the government have to subsidise the local car industry even more than before? Will that not add more strain to the consumers and the Federal Budget?

This same predicament of the car industry can also apply to the service industries. Despite all the urgings over 12 long years that their tariff protection will be drastically cut under AFTA, the car industry continued to be complacent and comfortable in the thought that the government will help them out.

Will the service industries also expect the same protective treatment and so delay in becoming more internationally competitive?

It could be far worse as the government has not publicly stated the deadlines for the different service sectors to become fully competitive internationally.

The service sector that is made up of the banking, insurance, shipping, port, airlines and other services and especially the professionals like accountants, architects, doctors, engineers and lawyers, etc., could well be dragging their feet in preparing for greater international competition.

Rafidah suggested that the industry players "should begin to operate with a mindset that is sharply attuned to a competitive marketplace, rather than being comfortable with the protection and support of the government".

The Minister's frank assessment is indeed true of all sectors of Malaysian trade and industry. The pertinent question remains: how competitive are we when we benchmark ourselves internationally?

And it may be too late to "begin to operate", with a competitive mindset now. We should have developed that competitive mindset long ago. Our competitors have already had a major head start over us, with their own more competitive mindsets.

But we could try to catch up on the lost time, if we go about promoting sharp competition immediately. But where are the signs that we are moving fast in becoming competitive, when almost all our policies, institutions and systems are still relatively protected?

We have to make our whole education system highly competitive from day one in school. Those who are disadvantaged can be assisted, but they should be taught to stand on their own two feet as soon as they are able, rather than holding their hands right through life. The crutches must be removed as soon as they can walk!

Similarly, all professionals and staff and workers must be rewarded according to their performance and not get their compensation as automatically and as a matter of right.

Unless we take drastic action to become more competitive and less dependent on protection, we will not be able to achieve the goals of 2020 on target!

Thus Malaysia faces great odds—not only at home but more significantly from international threats and challenges. If the rich and powerful industrial countries do not see the wisdom of giving developing countries like Malaysia a fair deal in international trade and investment, it will be an excruciating uphill climb for all developing countries in their advance towards developed-nation status in the future!

Proton Holdings is a case in point. Its net profit fell by a huge 54 per cent to RM510 million for the year ended March 31, 2004! Will this loss trend continue with greater global competition? But Proton claims that "for the current financial year, the demand in the domestic market has started to increase, based on orders received from its distributors"!

How is this possible, unless Proton has been able to quickly become more competitive against foreign cars on the basis of price and quality? Perhaps Bursa Malaysia should check these quarterly company statements to Bursa Malaysia more thoroughly, so that good corporate governance is more carefully and independently scrutinised in the public investors' interest! But the World Bank is no better!

The *World Bank Report* Condemns Rich Nations

The World Bank Report at its Spring meeting in Washington on April 14, 2003, condemned the rich countries for their bad trade policies that caused so much of the poverty in the poor developing countries.

The abysmal aid contributions of well below 0.5 per cent of their GDP, added to the feelings of deprivation and frustration and hopelessness in the poorest developing countries, have encouraged international terrorism to grow! The World Bank's chief economist Nick Stern boldly called the rich countries' trade barriers "absolutely outrageous" and stated that the "developing countries are right to feel somewhat aggrieved at the lack of progress", in the reduction of trade barriers and declining aid by the rich countries.

However, Stern's view that developing countries "feel somewhat aggrieved" is a serious understatement!

The developing countries, particularly the poorest agricultural nations, are in fact deeply resentful at the heavy subsidies given to the relatively rich farmers in the industrial countries. These unfair and suppressive agricultural subsidies are stifling the incomes of poor farmers in the developing countries. *This economic oppression alone can become a major source of frustration and can breed far more international terrorism in the future!*

Many developing countries are given no chance at all to provide food and shelter for their poor, because of the excessive greed and selfishness of some major rich countries, who are seen to exploit the poor countries.

So what alternative could the poor have for legitimate redress but for many to unfortunately resort to international terrorism?

Thus it is vital for world peace and security that the rich countries resolve to give at least the poorest developing countries fairer trade deals, even if they have to cut down on their aid.

The U.S. and U.K. that have dramatically shown their determination to start the war of so-called "Liberation" of Iraq (I believe it is for the oil) should instead take the lead in the current WTO negotiations, to liberalise the market access to the rich countries, for the poor countries.

This single initiative alone would greatly reduce the wide spread hunger and starvation and misery experienced as a human tragedy, in the poorest countries of the world.

As the German Economic Cooperation and Development Minister, Heidemarie Wieczorek-Zeul correctly stated at the Spring World Bank-IMF meeting in Washington in April 2003, pre-emptive wars are illegal, and that *"If there is a just war to be fought, it is the war on poverty and hunger and illness and disease, illiteracy and environmental degradation, exclusion and injustice"*.

Hopefully other rich countries will increasingly adopt that noble stand, to fight the root causes of terrorism!

This new realisation to show more compassion for peace and less cruelty in war, will help shape a far better New World Order. This aim will defeat international terrorism and improve the welfare of all humankind, especially those in poor developing countries—and not only for those in the rich and the powerful industrial countries of the world!

Therefore, in the future, the new war against international terrorism must incorporate strategies to fight the unfair and exploitative trade tactics of the rich and powerful countries against the poor countries. Only then will we be able to reduce wars and international terrorism—and win world peace and prosperity for all, and not only for some human beings in the rich countries!

Similarly we have to ensure that we are able to limit the adverse effects of disenchantment, disillusion and disunity among our own people and especially those who feel a sense of alienation among our multiracial society.

For this reason the Malaysian public service has to become increasingly more multiracial to reflect the racial composition of our Malaysian country, and to serve our multiracial society with greater justice.

The Recruitment of Non-Malays in the Public Service

The government has constantly expressed disappointment at the poor response of non-Malay applicants to join the public service. Deputy Home Minister Chor Chee Heung said at the 196th Police Day celebrations that "the number of Chinese applications was still declining".

At the same time, the Federal deputy management director (administration) Senior Assistant Commissioner Ahmad Hassan said that "the salary scale of the police force was the most serious deterrent".

A similar assessment was made by the Deputy Education Minister Dato' Abdul Aziz Shamsuddin when he addressed 800 teachers in Pekan, Pahang, on April 18. He said that "not many Chinese in the country want to be teachers or join the police and the armed forces as they prefer to be self-employed or venture into business."

These points are plausible but it is doubtful whether these are the only reasons for the poor recruitment of Chinese and Indians into the teaching profession, the police, the armed forces and indeed the whole public service, which is about a million strong! Dato' Abdul Aziz rightly pointed out that this ethnic imbalance in the public services "would not be healthy for the people and the country"!

Thus we need to undertake an independent and professional public survey to ascertain all the reasons for the low recruitment and the poor retention of non-*Bumiputeras* in our public services.

One major disincentive is the perception of unequal promotion prospects for non-*Bumiputeras* in the public service. The other common perception is that most recruiters generally do not show impartiality in the recruitment process!

Therefore one important solution would be to introduce recruitment quotas for the different Malaysian ethnic groups, along the lines suggested by Dato' Abdul Aziz for teachers that is, 55 per cent Malays, 35 per cent Chinese and 10 per cent for other races, or some such formula.

This strategy would reflect the ethnic composition of our country and should go a long way towards strengthening national unity, which, after all, is our overriding national goal.

In fairness though, there is now a growing realisation on the part of the government that much more should be done to encourage a greater sense of loyalty and patriotism among all races, and especially among the poor and the marginalised groups who may feel alienated by the government's development policies and especially the implementation of these policies.

A good example of the government's new policies to create a stronger sense of national identity and national unity is to be found in the policies and practices of the National Higher Education Fund Corporation (NHEFC).

The National Higher Education Fund Corporation (NHEFC)

The NHEFC was established in 1997 and is chaired by the Member of Parliament for Tambun in Perak, the innovative Dato' Ahmad Husni Hanadzlah.

The Fund has been increased by RM500 million to RM2.0 billion in 2003 from RM1.5 billion in 2002 and has already disbursed RM5.0 billion to 401,000 needy students, since it started operations only about 6 years ago!

However, because of its tight guidelines, many students were not able to avail themselves of its financial facilities.

Now the Fund is available to Malaysian students in private institutions of higher learning as well, unlike in the past when loans were restricted to only students in the government's institutions of higher learning.

The Fund has been liberalised further by making its Special Loan Scheme available to students whose parents have difficulties in paying their children's study fees, although they have a steady income.

In the past, these loans were provided only to the poor, the low-income groups and those who had uncertain sources of income, like the underemployed and the underemployed.

Hence many more poor students will now be able to benefit from the NHEFC than before. More non-*Bumiputera* students who attend private colleges and private universities will now have access to these subsidised study loans which were denied to them earlier.

This unfortunate situation had caused much dissatisfaction, disgruntlement and disharmony among the non-*Bumiputeras* in the past. Now this major source of friction and disunity has been removed!

The student loans provided are also more reasonable. The loans range from RM2,500 to RM5,000 for diploma programmes, RM3,000 to RM6,500 for first-degree Arts courses, and RM3,500 to RM7,000 for first-degree Science courses. They are quite adequate to cover the college or university fees and some other expenses. But poor students do get full loans to cover all their expenses, depending on the merits of the case, as determined by the NHEFC.

The Corporation plans to raise the income ceiling for the eligibility of these student loans. This will increase the accessibility to study loans for the more needy and bright poor Malaysian students regardless of race.

These policy changes will certainly help to strengthen national unity, especially for the young Malaysians who will feel marginalised and alienated, if they do not get the opportunity to pursue higher education!

More opportunities for higher education would be necessary not only to stimulate higher value-added output, but to accelerate economic growth. For this to be achieved, more economic stimulation is necessary.

The New Stimulus Package:
Not Up to Expectations

The much-anticipated Stimulus Package that was to be announced on April 7, 2003 has been further delayed! Was the delay necessary only because, according to Rafidah Aziz, they are still gathering data and inputs from all ministries and agencies!

But the package was delayed, according to earlier press reports on April 5, because "there was fine tuning of the new Stimulus Package—to ensure maximum effectiveness of the package"!

Then former Minister Mustapa Mohamed came out with a credible statement that the new package had to be updated to take into account the Severe Acute Respiratory Syndrome (SARS) scourge, which poses new threats and challenges to the economy.

Hence it is important to give a concerted and coordinated government view of economic issues and responses of this vital nature, as otherwise there will be more confusion and business uncertainty!

It is unfortunate that the package has been delayed as time is of the essence to sustain business and consumer confidence at this time of great uncertainty.

The present business distress is due to the war in Iraq and particularly SARS. However, the war which had been expected has taken longer than anticipated.

But the SARS epidemic, which was unexpected, is far more damaging to Asian economies and to the Malaysian economy as well. It is scary and is causing much anxiety!

Already the SARS situation is causing much business uncertainty and even public alarm! There have been 160 SARS cases to date (up to April 18, 2003) and 3,500 people worldwide have been affected by SARS for which, unfortunately, no antidote has yet been found!

Thus the U.N.'s Economic and Social Commission for Asia and the Pacific (ESCAP) cut down its economic forecast for developing countries in this region for 2003, to 5 per cent from its earlier estimate of 5.4 per cent, due to the impact of the SARS, in April 2003.

If, however, the SARS spreads further and gets more out of control, the growth rates will have to be reduced even more.

On the other hand, the Organisation for Economic Cooperation and Development (OECD) estimated in April 2003, a lower growth of only 1.9 per cent for 2003 and a 3 per cent expansion in 2004 for its 30-member rich countries.

Thus the Stimulus Package could have been introduced earlier and then supplemented with more proposals later if necessary, particularly since the "package has been almost finalised"!

The Stimulus Package which could be announced in May need not be confined to additional fiscal measures like more expenditure allocations and more tax cuts. Indeed there is not much we can afford to give away, since the Budget has been in deficit for the last five years.

Both the Acting Prime Minister Abdullah's useful Dialogue with about 120 corporate leaders and Minister of

International Trade and Industry Rafidah's Annual Dialogue with leaders from about 118 business associations, which were held separately on April 22, stressed the need for the government to reduce the cost of doing business in Malaysia.

The businessmen asked for reductions in electricity and water, and sewerage charges, corporate taxes, and for the provision of soft loans and even subsidies to overcome the adverse effects brought about by SARS.

However, I was surprised that no one mentioned how the Budget could bear this additional burden without hurting the economy as a whole!

Nevertheless, there were some convincing arguments made to review: the Foreign Investment Committee (FIC), the Industrial Coordination Act, the slow implementation of the productivity-linked wage system, the abolition of the Automatic Pricing Mechanism (APM) and the AP (Approved Permit) control on steel products and the requests to cut Employees Provident Fund (EPF) contributions!

Much concern was also expressed by the American Malaysian Chamber of Commerce (AMCHAM) that "Malaysia has lost a number of investments because its government agencies have not been effective in attracting large companies". It mentioned that public seminars alone will not finalise investments!

AMCHAM therefore called for the setting up of a "One-Stop Foreign Direct Investment Centre". This reflects the frustration experienced by AMCHAM and other foreign investors over so many outstanding issues. These relate to immigration, greater flexibility in equity structure requirements and reducing and simplifying the application and the long processing time for pioneer status!

It is very surprising that these issues continue to bug American investors. Why can't our government agencies overcome these problems once and for all. If we cannot or do not want to accept the proposals of these foreign investors, we should tell them that we are not prepared to change our policies and accept the consequences of declining FDIs.

There is just no point in carrying on with government's Annual Dialogues and expensive international investment road shows, if at the end of it all, we are told that our agencies have not been effective!

Indeed government should respond to these serious criticisms publicly and decide on the AMCHAM proposal to establish a one-stop agency as a matter of priority. We cannot afford to lose out on FDIs at this time of greater competition under globalisation!

The underlying theme is to increase transparency, to speed up decision-making and overall administrative efficiency.

Indeed of the 287 trade and investment issues that have been raised at the MITI Dialogue, 165 are "repeated issues"!

How can we step up economic growth if so many important business issues mentioned above, have not been resolved for so long?

MITI Dialogue for 2003 was reduced to 3 days from the 5 days of the Dialogue in 2002. Rafidah explained that less time was necessary as the communications between the government and the private sector had improved.

But this change does not throw light on why so many outstanding issues are repeat issues which have not been solved? On the contrary is the shorter MITI Dialogue due to the need to avoid the embarrassment caused by repeated rejections of the requests to change some trade and investment policies? This is by no means clear and should be clarified to increase business confidence!

Furthermore, Prime Minister Dr Mahathir Mohamad and now Acting Prime Minister Abdullah Ahmad Badawi have repeatedly urged, that there is an urgent need to change mindsets—to reduce bureaucracy, to raise the quality and productivity in the public services to introduce greater facilitation of business and foreign investment, and to minimise the "dependency syndrome".

But how much of our policies and thinking have really changed? The government could facilitate the private sector to overcome the present difficulties associated with SARS and the war in Iraq, more effectively, by responding positively to the private sector's requests for policy changes, instead of straining the Budget too much, by cutting taxes, increasing expenditures and providing subsidies to industry!

Hence the Stimulus Package need to have been brought forward more purposefully, to counter the adverse effects of the war in Iraq earlier than later. The Stimulus Package should not turn out to be a case of too little too late?

The economy needs stimulation at the right time, otherwise it can weaken and decline further!

The Acting Prime Minister Abdullah Ahmad Badawi stated outside the Parliament chambers on April 7, 2003, that "If we make the proposal [Stimulus Package] early, we may have to change it, as the war situation might warrant the changes".

April 2003 is not really too early to introduce the Stimulus Package, especially when the public have been waiting for over a month for the promised package.

Furthermore, it is not wrong to introduce the package now and then introduce other proposals later if necessary. After all the economic situation is very dynamic and no time is perfect for the required changes.

Already the government is considering whether to reduce the scope of the projects in order to cut down on costs or carry out the projects under the 8th Malaysia Plan in phases—according to Deputy Minister of Public Works Mohamed Khalid Nordin.

Indeed the mindset changes can and should be introduced as soon as possible to counter the economic slowdown.

Hence it is unfortunate that the Stimulus Package has been further postponed. I hope there is not too much undue delay, as the economy will then suffer even more.

But now the question increasingly rises as to whether the Stimulus Package will be strong enough to counter the depressing effects of the nearly one-month-old U.S.-led invasion against Iraq, since it started on February 20, 2003?

Worse still, will the Stimulus Package be able to pack enough punch, to counter the regressive effects of the SARS attack on Malaysia and other parts of the world?

Already the Malaysian Institute of Economic Research (MIER) has cut its earlier growth estimates of the Malaysian economy from 5.7 per cent made in November 2002, to 3.7 per cent as at April 2003

Although Dato' Mustapa Mohamed, the executive director of the NEAC, has maintained that "The country's fundamentals are still strong and the uncertainties are just a temporary aberration", he may be somewhat optimistic!

While it is true that Malaysia's economic fundamentals are still strong, it is well to bear in mind that strong fundamentals can be weakened by unexpected extraneous factors like the war in Iraq and now the dangerous SARS scourge! And strong fundamentals can soften very fast if the threats to our socioeconomic well-being and business and consumer confidence are undermined for too long.

It is therefore better to be on guard and cautious and to take pre-emptive action early enough to protect the economy against the challenges to our strong fundamentals.

Given these difficult developments, would Prime Minister Dr Mahathir have to push forward further reforms to strengthen the socioeconomic and political structures, before he hands over the reigns of to Abdullah Ahmad Badawi before he retires in October 2003?

Can the Prime Minister-designate Abdullah Ahmad Badawi prevent economic decline, before he takes over the leadership of the country from Dr Mahathir in October 2003?

We have to liberalise, be more responsive to the private sector's valid requests and become far more competitive if we are to overcome the serious challenges from SARS and the war in Iraq.

We hope the Stimulus Package provides some of the most important solutions soon!

Dr Mahathir has already warned about the dire consequences of the war in Iraq on the world economy. And if his prognosis comes to pass, then Malaysia will be forced to face even greater socioeconomic problems!

The package of new strategies that was finally announced by the National Economic Action Council (NEAC) on May 22, 2003, was sound and stimulative. It contained an impressive record of 90 specific measures to boost the economy and to strengthen its fundamentals.

The Package was innovative in that although the total increase in the allocation of funds amounted to a massive RM7.3 billion, only RM1.7 billion was incurred by the already strained Federal Budget.

The rest of the RM5.6 billion was provided by the financial institutions. Thus the deficit Budget was spared too much undue strain.

This financing technique has been quite unique. It has not been undertaken on such a large scale in the country before.

The question that now arises is, why this technique was not adopted so widely before. Is it because the reserves of these financial institutions were too jealously guarded? Were these funds committed but not really loaned out, because of conservative lending policies? What has changed that will enable these new funds (RM2.0 billion from Bank Negara and RM3.6 billion from other major government financial institutions) to be fully disbursed as loans now?

These financial institutions must be monitored more closely by the government to get them to lend more aggressively. Bank Negara has also to ensure that the funds are properly and prudently lent out and recovered by these financial institutions, so as to prevent wastage and rising non-performing loans.

The large funding is impressive. However, the Stimulus Package could have achieved much more, if the opportunity was taken under the current economic crisis, to restructure the economy more purposefully.

For instance, there were expectations that the Foreign Investment Committee and the Industrial Coordination Act, would be removed or more significantly revised. But the ICA remains and the FIC guidelines were not changed substantially enough to attract foreign investors away from the more competitive neighbouring countries.

Perhaps the package was designed to overcome the immediate challenges posed by SARS and the U.S. invasion of Iraq, before we take more proactive measures to overcome the medium- and long-term structural problems facing the economy. These challenges would include the relatively low productivity and weak competitiveness in the economy.

Nevertheless, the package was better than expected in its scope. It was indeed the biggest Budget between Budgets!

The Stimulus Package aimed to reduce Malaysia's dependence on the external sectors by:

1. Promoting private investment;
2. Strengthening competitiveness;
3. Developing new sources of growth; and
4. Enhancing the effectiveness of the delivery system.

Firstly, promoting private investment through the development of SMIs as a catalyst of growth is laudable. However, Bank Negara will have to adopt more punitive measures against those financial institutions that continue to be indifferent to meeting the micro-financing targets. Otherwise, SMI financing will continue to suffer!

It is gratifying that the *Bumiputera* equity participation will now be applied consistently by all ministries, except where exemptions are granted by the government. Too many 'exemptions' will negate this good policy!

This important measure will remove the uncertainty and inconsistencies that occurred previously, much to the frustration of private investors. However, it would be useful to employ more transparent criteria for the granting of exemptions to the general 30-per-cent *Bumiputera* equity ownership policy.

Secondly, strengthening national competitiveness has encouragingly been given high priority.

The increased fiscal incentives given to R&D is welcome. But it is important to ensure that the tax incentives are speedily approved and that the best research projects and the most qualified researchers are provided the financial resources.

It will be a pity if some of our research brains are discouraged from utilising these attractive research funds if they are excluded on ethnic grounds!

The provision of an additional RM500 million to the Skills Training Fund as loans to poor students for technical courses, will no doubt help increase the output of much needed technicians. *Hopefully the terms of these loans will be attractive enough for the poorer students, regardless of race.*

The establishment of the RM100 million Retraining Fund for new graduates for re skilling in selected fields such as IT, begs the question as to why these graduates need to be retrained in the first place.

Are the universities turning out unemployable graduates and if so why can't the universities be restructured themselves? It may be more prudent to tackle the problem of unemployed graduates at the source of the problems—the schools and the universities!

The greater flexibility that will be given to employers in hiring highly skilled foreign workers, will be accepted with great relief. Although we want to attract higher value-added foreign investment, some of our government agencies do not seem to appreciate the importance of giving speedy approvals for high value-added workers.

Many Malaysian and foreign investors have been so frustrated by the bureaucracy and delays that they have moved out especially to countries which are more flexible on the employment of foreigners.

One way of discouraging foreign investment is to be excessively bureaucratic and slow in decision-making or being corrupt!

Thirdly, the development of new sources of growth will surely enable Malaysia to mak e a breakthrough in restructuring our economy to better meet the new challenges of globalisation.

Hence our relations with multinationals and major trading partners like Singapore are important.

CHAPTER 6

MULTINATIONALS AND SINGAPORE ISSUES

MULTINATIONALS in Malaysia and other countries need to consider their role in balancing their legitimate quest for profits and their obligation to fulfill their social responsibility to their host country and society. This is particularly relevant to Third World countries.

Firstly, let me state that Malaysian multinationals can feel satisfied with their long record of achievements in Malaysia.

Secondly, my question, however, is: are multinationals satisfied that they have achieved the right balance between profits and social responsibility?

Are the top international performers among our multinationals and what are their strengths and weaknesses in this balance?

The answer depends how high the standards of multinationals are? Do they want to be just good corporate citizens or outstanding multinationals that can be proud of being exemplary?

It is so easy for multinationals in Malaysia to be self-satisfied and to become complacent with their success.

It is also comfortable to judge the performance of multinationals by the number of good governance awards that they win. But they must remember that it is more difficult for them to judge their performance through the eyes of others who work outside the multinationals!

Do multinationals in Malaysia want to be exemplary and, if so, how special or different are Malaysian multinationals compared to others? Do multinationals serve the social needs only at the micro level and in just narrow micro ways?

What proportion of their profits do multinationals invest to fulfil their social responsibility and is the proportion up to best international practices? Can we propose that multinationals have a target of 1 per cent of profits to be allocated for community services? (This goal would be consistent with the U.N. Resolution for the rich industrial countries to provide about 1 per cent of their GNP as aid to developing countries.)

Let's examine these issues.

This topic of balancing profit with social responsibility is intriguing because it has both national and global concerns. The subject relates to both good international corporate governance and also reflects the NEP, as the policy was intended!

Balancing profit and social responsibility gives rise to, challenges and threats on two fronts, viz:

1. Globalisation challenges, and
2. Domestic socioeconomic policies and practices.

Globalisation Challenges

1. Multinationals need to realise that, rightly or wrongly, they are often regarded as wolves in sheep's clothing!

2. Why? Because of perceptions that they are foreign owned, staffed at the top usually by foreigners and operate for self-interests, making large profits that are repatriated continuously.

3. The globalisation process has been seen as evidence that many rich and powerful countries and governments would use their multinationals as instruments to further their foreign policy objectives, to spread hegemony and dominate the globe! This is seen particularly in the case of the U.S. and the U.K., especially after the Anglo-American invasion of Iraq!

4. The WTO negotiations that are held in "Green Rooms" add credence to this belief. When the WTO pushes headlong for rapid liberalisation and "opening-up" measures to expand trade and investment, the question often asked is: for whose main benefit? It is not altruistic and not intended to benefit the host country but those of the foreign shareholders.

5. When the pursuit of "unrestricted profit" to please shareholders is the goal, how is it possible to build confidence that multinationals are bona fide? How is it feasible to accept that multinationals are not interested in profit, first and foremost, and social justice (including the protection of the environment) as an afterthought?

6. These perceptions can be summed up thus: that multinationals are often regarded as a necessary evil

that poor countries need in order to move up the
ladder of economic development because of the
technology, training, employment and foreign
markets that accompany multinationals!

Domestic Challenges

Multinationals constantly face the following challenging
questions in Malaysia and indeed elsewhere too:

1. Do multinationals in Malaysia merely seek to
   optimise profits and only pay lip service by meeting
   only the basic minimum "social commitments" to
   the Malaysian society?
2. Do multinationals undertake to attain sustainable
   development, more for the sake of presentation, to
   show a nice face?
3. Do they try to sustain development as a whole or do
   they want to actually "promote sustainable
   development projects" to create the image of being
   proactive.
4. Do multinationals identify themselves even more
   with the dynamics of growth and distribution
   aspects of Malaysia's NEP and Vision 2020
   aspirations or are they merely selfish and
   self-centred?

The promotion of sustainable development goes beyond
the narrow confines of pursuing excellence in the company
alone. It implies not only setting themselves up as
benchmarks for good governance, but also to help spread
goodwill and to share in their profits. Multinationals need to
encourage more investment, more transfer of technology

and the participation of their senior officials in dialogues and policy formulation with the government.

Malaysian multinationals should not wait to be called to serve and to use their great international resources to provide solutions to the many challenges and problems facing Malaysia, especially at the macro policy levels.

What are the Challenges for Multinationals in Malaysia?

1. **To show that multinationals in Malaysia are different, limit the dividends from profits at a reasonable level, say 15-20 per cent per annum.** This would balance the aspirations of shareholders and that of the host countries. Then there would be greater confidence that multinationals are conscious of the goals of justice and equity and the proper balance between profit and social responsibility! The excess profits can be invested locally.

2. **Be more transparent by honestly showing scorecards that indicate the total value of their contributions for social causes, in comparison with international norms.** Like the U.N.'s aid target for industrial countries, is there such an index for multinationals. Can Malaysian multinationals start such a sustainable development index?

   This will highlight transparency and accountability. Otherwise, we only read of the commitments made by many multinationals for fulfilling their social responsibility in Malaysia, without being able to benchmark their contribution to sustainable development.

   However, we find that issues like "Do no harm to people", "Use energy efficiently", "Promoting best practices", etc., are mostly aimed at optimising the productivity and profits of the multinationals, and are not directly intended to serve society at large *per se*!

So, is there a proper balance between seeking profits and serving the multinationals' social responsibility in Malaysia? One of the most important commitments would be to reduce the Digital Divide like Internet Desa which is about teaching rural folk to use the Internet. It is this kind of project that is more worthy than the small *Hari Raya*, Chinese New Year and Deepavali parties and children's outings that are organised by some multinationals!

Have multinationals carefully assessed the public perceptions of them in Malaysia? Could some of these perceptions be along the following lines:

1. Multinationals in Malaysia are colonial in nature.
2. They are modern and highly technological, but not so keen to fully share their knowledge with Malaysians.
3. They have a relatively low profile and one wonders why. Have they got matters to hide?
4. Malaysian multinationals have made and repatriated immense profits over many years in Malaysia, some from well before World War II?
5. Multinationals do not have enough local content in staff and tend to keep the top jobs for themselves, although there are many suitable Malaysians, who can do their jobs.
6. Multinationals are small on charity, because they are too preoccupied with maximising dividends and profits for their majority Western shareholders!

The Way Ahead for Multinationals:
Some Proposals

Do get involved in macro and national issues by quietly offering your help. *I can think readily of several critical areas where international intellectual resources of multinationals can be utilised to benefit Malaysia, as follows:*

1. Share long-term strategic world outlook scenarios with Malaysian planners.
2. Increase productivity in the civil service by sharing some of the latest human resource technologies.
3. Encourage more R&D in Malaysia through mechanisms that strengthen university and industry collaboration.
4. Provide more scholarships and training in schools and universities.
5. Extend Best Practices of the multinationals to Malaysian SMIs.
6. Participate more actively to support the development of good corporate governance.
7. Establish libraries or other centres of learning, especially in the poorer sections of the Malaysian community in both urban and rural areas of the country.

Then multinationals will show that they are different and that they really want to be caring, distinctive and different from all other multinationals—and it won't cost much more! Then Malaysian multinationals will truly be better able to improve their balance between profits and social responsibility in Malaysia!

Multinationals have done well in Malaysia, but they can do much better!

I would suggest that multinationals in Malaysia look into the following:

1. Study and revise their commitment to promoting sustainable development in Malaysia.
2. Harness all their available resources to work on new priorities to strengthen economic development.
3. Expand the public's awareness of their many good contributions by reaching out more effectively to the needy.
4. Level up their micro-social commitments and contributions to the higher macro levels.
5. Lessen their feelings of complacency and of having done their share in contributing to sustainable development.

Finally, can multinationals in Malaysia lead other global companies to pledge 1 per cent of their profits to community services in Malaysia, for other multinationals to follow? How much more will multinationals be able to contribute to the Malaysian economy?

Multinationals have to counter negative perceptions to prosper. Free-trade agreements that are fair could help multinationals improve their image and investment prospects.

Singapore Issues
The U.S.-Singapore FTA Poses Challenges to Malaysia

The Singapore-U.S. Free Trade Agreement (FTA) that was signed between the world's biggest and perhaps smallest industrial countries on May 8, 2003 will have significant implications on future trade and investment in Asean countries and in Asia.

It is the first FTA between the U.S. and an Asian country. It is a precedent and prototype that other Asian countries would want to follow, to fight competition from this FTA.

What the U.S. could not win in its negotiations in the WTO, the Singaporeans have speedily given to the Americans, despite consideration of any disruption it could have on Asian economies.

The strategy of gradual and acceptable opening up of the Asian markets to the superpower with superior competitive capacity, will therefore now be upset.

The U.S. will use Singapore like a Trojan Horse to smuggle a faster pace of so-called free trade into the Asian continent, regardless of whether Asia is ready!

The service industries in Singapore will be dominated by U.S. service industries like banking, insurance, shipping health and education and technology. Singapore could thus become a foreign metropolis while larger and longer established countries would want to protect their national sovereignty.

The U.S. service industries will now be able to make deeper inroads into Malaysia and other Asian countries that are not prepared to open their economies as fast as Singapore, in order to protect their economic integrity.

It will be ill advised to assume that the U.S.-Singapore FTA will not adversely affect Malaysia and the region. From now it will be even more likely that banking, insurance and also professional services will be made more readily available to Malaysian companies and individuals by American and Singapore-American enterprises in Singapore! They will just visit Malaysia on business or tourist visas and take away Malaysian clients and businesses, by taking advantage of the proximity of Singapore to Malaysia and Asean.

Thus international competition will increase at our very door, which the U.S.-Singapore FTA will pry open the Asean market, to their advantage. Many of our professional and businessmen, not used to sharp competition so close to home, will lose out.

Singapore will be able to further consolidate and expand its role as the regional service and especially financial hub. Malaysia's aim to become such a hub would then be weakened.

What should Malaysia's response be? The earlier goal of moving forward gradually in order not to disrupt the steady development of the domestic and regional service industries, will now have to be reviewed and changed.

Liberalisation would now need to move at a faster pace. This may cause some painful structural business adjustments, which Asian countries were generally wanting to avoid.

After signing the FTA with Singapore, the U.S. would be encouraged to push other Asian countries to go their separate ways and to sign more FTAs with America. *Malaysia will then have to follow suit or risk being left out or marginalised.*

This would be easy for the U.S. to do as the Americans are already censuring and wanting to impose sanctions and 'punish' countries that did not support the U.S.-led coalition's invasion and the military occupation of Iraq.

Malaysia and Asean will have to be more wary of the U.S. agenda in Asia especially after Senior Minister Lee Kuan Yew's recent statement that "without the U.S., terrorism cannot be stopped"!

Indeed, Singapore, having readily made available its military facilities to the U.S. and signing the FTA with them, it could appear to become a U.S. outpost in Asia. This could hurt Asean and other regional interests.

Hence now that Singapore has signed the FTA with the U.S., Malaysia and other Asean countries would have to review and even revise their trade policies with the U.S. and Singapore as a matter of priority.

Malaysia may now want to consider establishing its own FTA with the U.S. as part of its overall policies to become more internationally competitive, in accordance with the government's package of new strategies announced on May 22, 2003.

The U.S.-Singapore FTA will Erode Asean

The rapidly expanding free-trade agreements (FTAs) around the region have to be watched carefully and solutions found to counter any detrimental effects on Malaysia. Let's examine, for instance, the U.S.-Singapore FTA.

The U.S.-Singapore Free Trade Agreement which came into effect on January 1, 2004, could undermine the Asean Free Trade Area (AFTA) and even Asean itself.

Now all U.S. goods entering Singapore are duty free. Nearly 80 per cent of Singapore's exports to the U.S. will enjoy duty free status and this percentage is expected to rise to 92 per cent in just 4 years' time!

This closer collaboration between Singapore and the U.S. will thus gradually force other Asean countries to also follow suit. Then Asean will no longer have an exclusive Free Trade Area, to first build its economic resilience and competitive muscle, before it can collectively take on the rest of the world.

The implications are that the industrial countries like the U.S. will overwhelm our weaker developing economies of Asean and gradually dominate them!

Already there are some 1,300 American companies and 15,000 U.S. nationals with a total investment of US$27 billion in the tiny island of Singapore. How much more U.S. and other foreign control can Singapore take before its economy is taken over by U.S. and other economic interests?

To what extent can Malaysia and other Asean countries counter the invasion of U.S. goods and services that could come in, through the back door of Singapore? U.S. and Singapore businessmen could claim that there is a 40 per cent Singapore content in its exports to Malaysia and other Asean countries!

In the service sector, U.S. banks, insurance companies and legal firms will now be open to U.S. participation. This means that U.S. companies could get hold of financial and legal business in Malaysia and other Asean countries, using Singapore as a base for their operations.

Henceforth, hundreds of foreign companies from Europe, Japan, Australia and New Zealand and other potential Singapore FTA partners will be will be able to capture Asean business via their control of Singapore companies. Hence our Malaysian and other Asean businessmen and professionals, will need to be publicly advised by our authorities of the serious threats they face, from the Singapore FTAs with the U.S. and other industrial countries. Then they we will be better prepared to take on these major new global challenges against developing Asean countries.

Asean countries will need to develop new national trade policies and work more closely together to safeguard their individual and collective integrity against foreign economic forces. As we develop more FTAs with other countries, we also need to monitor our trade and economic relations with the great multinational companies that could dominate our economy.

Singapore's *Kiasu* Mentality
It is now clear to most Malaysians that Singapore appears to have a real thirst for "oneupmanship" that may result in the island city state going thirsty!

The majority of Malaysians were shocked to read Singapore's Foreign Minister S. Jeyakumar's reported statement on October 31, that Malaysia has no legal right to a price review of water, since it should have been done before the water agreements expired in 1986 and 1987!

Could Jeyakumar be accurately reflecting the views of Singapore? If so, can the Singapore authorities be serious about taking such a preposterous stand on the pricing of Malaysia's water supply to Singapore?

While it is true that the pricing should have been negotiated in 1987, does it mean that just because Malaysia did not rush into price negotiations, because of other priorities, that Singapore should be so legalistic and take advantage of the situation?

I can understand our Malaysian officials taking the fraternal view that, given our special and close historical ties with Singapore, we need not be so legalistic and insensitive as to start negotiations immediately after the expiry of the water agreements!

But I suppose our Malaysian officials were wrong in assuming that we were dealing with a close and sincere neighbour. We did not think that they would insist on little legalities, rather than the real spirit behind the law and to act as a good neighbour.

But now we know better! From now on, Malaysia should treat Singapore very differently in all our economic, business and other relations. In fact, we need to classify all our trading partners according to new criteria that will be based on the degree of sincerity, goodwill and the willingness to work together for mutual benefit.

Countries that are regarded as unfriendly should be regarded as unreliable partners who will be waiting to stab us in the back, and thus should be treated accordingly.

It is also important to ascertain early, as to whether we are obliged to adhere to the provisions of an intrinsically outdated and bad water agreement. Is it right in our peoples interests, to sell our precious water at the ridiculous price of 1.5 Singapore cents for 1,000 gallons? No legal document that is indefensible or even devious, should be upheld for the sake of mere formalities.

Hence the door should be left open for further negotiations on a fair and reasonable market price for supplying water to Singapore to meet its desperate need for drinking water.

But if the price negotiations fail again, we should then go for arbitration, although this could be the last straw!

I sincerely hope that this vital water issue should be settled soon and within a definite time frame. Otherwise, irreparable damage will be done to the future relations between our two countries, that ironically were just one country, not so long ago!

Unfortunately, this thorn in the flesh for both Malaysia and Singapore, has not been resolved by Prime Minister Mahathir.

Hence, it is now left to Abdullah Ahmad Badawi to settle this long outstanding issue, as early as possible, with dignity and pride and pragmatism, on the part of both sides.

Then both our economies will be able to better integrate and prosper at a faster pace in the future.

There are obviously some hidden factors that have not been brought to the public notice, that underlie the problems Singapore faces in settling the price of water.

Singapore's Economic Slowdown
One reason why Singapore is being so difficult in paying a reasonable price for the water that it gets almost free from Malaysia is its present economic stress in 2002.

It is because Singapore's economy is on a slippery slowdown that it just cannot afford to pay a more reasonable price than the mere Singapore 1.5 cents per 1,000 gallons for water from Malaysia!

It is as simple and as bad as that!

So Singapore's strategy may be to be unduly tough in its negotiations and to stick to the letter of the law in the old water agreement, rather than to observe the spirit behind it. That is Singapore's shortsightedness and failing!

The latest report issued for the third quarter of 2002 gives a dismal economic scenario for Singapore.

Its growth rate for 2002 is now estimated to drop from the earlier estimate of 3.0 per cent to only 2.5 per cent for the whole of 2002!

Friedrich Wu of the Ministry of Trade and Industry in Singapore told a press conference recently that Singapore did not rule out even a "double-dip recession" for 2002!

Already several big multinationals like Daimaru have moved out of Singapore after 19 years and Agilent Technologies and Chartered Semiconductor Manufacturing have retrenched their staff significantly.

The Singapore government has characteristically set up another high-level Economic Review Committee to examine its economic structure that is highly dependent on exports because of its minuscule-island service-oriented economy.

But this Review Committee has already made its first blunder in reducing income tax rates and raising the sales tax from 3.0 per cent to 5.0 per cent. This will raise the cost of living and the cost of doing business in Singapore, although the authorities are also trying to restrain wage increases.

Thus with so much insecurity and disenchantment with Singapore's economic model, it can be difficult for the Singapore government to be fair and reasonable in its dealings with Malaysia on the bitter water pricing issue.

But Singapore should have a longer-term view in its analysis of its relations with its neighbours and its major trading partners. In being overly legalistic and niggardly on the pricing of vital water supplies from Malaysia, Singapore might win the battle but lose the war in the longer run! As a good neighbour lets hope Singapore's weak economy will recover soon. But in the meantime, we also hope that Singapore will see the need to pay its fair dues, for the sake of its national pride and better long-term relations with Malaysia.

But if Singapore continues to take an unreasonable stand and insists that Malaysia has no right to negotiate for a price review, then resorting to arbitration of the terms of the outdated 1927 Water Agreement appears to be the only answer.

It is therefore understandable that the Malaysian government has now taken the stand that there is no point in talking further with Singapore and that it would be better to resort to arbitration!

But if arbitration can be avoided by undertaking negotiations between two close neighbours, under the new leadership of Abdullah Ahmad Badawi after October 2003, that will be useful!

Then it will be all for the better for the future goodwill and close neighbourly ties between our two countries that are so inextricably connected, almost like Siamese twins!

Our relations can in fact be based on the exemplary ties that we have with Thailand, with whom we enjoy cordial and mutually beneficial friendship.

The quality of our economic relations with our neighbours in Asean will have a strong bearing on our own economic performance and progress in 2003.

CHAPTER 7

A REVIEW OF THE MALAYSIAN ECONOMY IN 2003

WILL the Malaysian economy move forward, stand still or slide?

The January 27, 2003 U.N. deadline for the *Blix Report* to the Security Council has passed—and the U.S. War against Iraq has not started yet?

What happens next in Iraq will determine the economic outlook for the world and Malaysia as well!

The U.N. Report on the World Economic Situation and Prospects 2003 has already lowered its economic growth estimates to 2.75 per cent from 2.9 per cent just three months ago.

It states that "an escalation into military action would have even more profound negative economic consequences"! Consumer and business confidence will then decline further, while petroleum prices could rise!

Deflation could then occur on a large scale all over the major economies.

In the U.S., overcapacity and oversupply, especially in the electronic and telecommunications industries, have dampened investment. Consumer spending has also been slow because of the fears of war in Iraq and the great concern that incomes and even peoples' jobs will suffer.

According to Reuters' Global Survey of major economies during January 8-15, deflation has troubled Japan for about three years!

For fiscal year 2003, the Japanese economy is expected to grow by only 0.9 per cent. Although Japan's exports have been performing well, the rising value of the Yen against the U.S. dollar, is adding further pressure to depress the Japanese economy with deflation.

The weak banking system in Japan, with its huge backlog of non-performing loans, has been eroding business and consumer confidence and holding back any real recovery.

The consequent adverse impact on the Malaysian economy could therefore be significant.

While we keep our fingers crossed about the international economic uncertainty, we must nevertheless soldier on, to build a higher degree of resilience on our domestic economic front.

Prime Minister Dr Mahathir commented after the opening of the new wing of the National Institute of Public Administration (INTAN, or *Institut Tadbiran Awam Negara*) that "We could do pump-priming to get more funds to circulate and spend money on projects"!

But can the economy accept even more budget deficits?

How much more debt accumulation can we afford? And how do we deal with the falling income from exports and the slowing down of economic growth?

Central Bank Governor Tan Sri Dato' Dr Zeti Akhtar Aziz rightly stated at the Global Forum in Putrajaya on January 9 that "If you look at the domestic environment, there is no point where you would consider it a pressure point". This is a fair assessment at the present time. But what of the future and what are the likely trends against which we have to be vigilant?

With the great deal of uncertainty worldwide, pressure points could develop in the economy, if not now then quite soon. It will be imprudent to just draw comfort at this time from the perceived lack of pressure points in the economy, when in fact they may be building up in the very near future!

The new economic Stimulus Package that is expected will be in addition to the two previous Stimulus Packages of RM3.0 billion in March 2001 and RM4.3 billion at the end of 2001. The total of RM7.3 billion was in addition to the Budget development expenditure of RM34.0 billion for the whole of 2001.

This additional spending resulted in the Federal Budget recording a deficit of RM18.4 billion in 2001 and RM16.7 billion in 2002. The budget deficit as a percentage of GDP is now estimated at about 4.7 per cent for 2002. But this deficit could worsen, with the new Stimulus Package coming soon!

Then we will also have to borrow more and increase the public debt!

The net development expenditure for 2003 is about RM33.0 billion. It includes some large projects like the electrified double-track railroad estimated at RM12.0 billion, the RM5.7 billion water pipeline between Pahang and Selangor, the Second Link Penang Bridge costing RM2.3 billion, and the RM1.5 billion Southern Customs complex.

Pump-priming the economy with huge investments in infrastructure projects is a major concern but equally important is to ensure that these projects are implemented with efficiency and cost effectiveness.

Many of these projects are not necessarily tendered but contracted out on a "negotiated tender" basis! This means that the prices could be higher than if the projects were offered on "open tender", where there is competition for which we could get the best price and quality from tender bids.

If we have to pay higher prices and at the same time get lower quality projects for these "negotiated tenders", there will be much wastage of public funds, and the economy will suffer in the long term.

Many of the tenderers who win the negotiated contracts are not those who will actually implement the contracts. Often enough these contracts are subcontracted out to those contractors who can actually perform. But they are "squeezed" on their profit margins. Thus these subcontractors "cut corners" in their building of infrastructure projects or in the supply of goods and services!

In the end, the tax paying public has to pay for all these non-productive activities and the public also loses out further by having to accept substandard and low quality projects.

Many of these bad contracts also have a high import content in the form of imported plant and machinery—and this could strain the balance of payments.

This is one major area of continuing concern which must be addressed with a stronger will and greater priority.

Perhaps the new Prime Minister will be able to place this important issue high up on his Agenda for Action when he takes over the reins of power towards the end of the year.

Already the balance of payments are weakening. Exports for November 2002 grew at a slower pace of 9.9 per cent as compared to November 2001, while imports increased by only 4.9 per cent as compared to 11.1 per cent for the same period in 2002. This indicates that the import of intermediate goods have declined and will adversely affect production and export of electrical and electronic goods in the near future!

The balance of payments whose surpluses have been narrowing, will also be further strained unless our service sector becomes more competitive. But our professional groups continue to be slow to open up to external competition. They cling to protection. So how can they learn to be internationally competitive?

Now there are new challenges. Singapore's Prime Minister Goh Chok Tong and Thailand's Premier Thaksin Shinawatra agreed, during an official meeting in Bangkok on January 12, to move towards a new concept of the "Asean Economic Community"!

They plan to achieve this status step by step, in about five to 15 years. They want to relate their Free Trade Area arrangements with the U.S. to Asean Integration and to provide the model for other Asean nations to follow.

This raises the sensitive question as to how Malaysia is going to respond to this new challenge posed jointly by its two immediate neighbours—Thailand and Singapore?

Malaysia could be caught in the middle! Both neighbours can liberalise and attract far more foreign investments and even pull investments currently now in Malaysia, to the more attractive liberal investment regimes of the two integrated Asean economies.

So far there has not been any comment from Malaysian policymakers on the implications of these economic policy initiatives by our closest neighbours.

These moves will make their economies more competitive than ours and we could lose out in the ensuing greater global competition especially in the service sector.

It is again impractical to act like ostriches and to hide our heads in the sand and thus ignore the rapid competitive developments that are taking place around us!

The answer to is to liberalise our service sector particularly the banking sector, at a faster pace.

The banks are once again being allowed to enjoy profit margins that are far too wide! The deposit rates are around 2.5 per cent, the Base Lending Rate (BLR) is about 6.5 per cent, but the actual bank lending rates are about 8-9 per cent even for solid customers.

Furthermore, some banks have become so averse to risks that they want to call back loans to customers with good track records in servicing their debts, just because the share prices of stocks used as collateral have weakened.

How efficient and proactive are the banks. Some are no better than moneylenders. Bank Negara should review the role of banks in complementing national economic strategies and stimulating economic growth. Banks that are not fulfilling reasonable expectations of them need to be consolidated soon with other more successful banks.

But we must have a whole new economic package to stimulate and sustain higher economic growth. The much-awaited new economic measures have not been announced by the NEAC as yet.

We cannot afford too much delay to counter the many uncertainties and the slow economic growth. Other countries are moving faster—hence we must measure our speed of change against their performance and not against our own internal standards. Our internal standards are often governed by the influence of our businessmen and professionals who clamour for more and more protection for longer periods of time!

The U.S. economy is facing slow growth and possible deflation. Thus President Bush announced on January 8, 2003, a 10-year tax cut package of US$674 billion, to stimulate consumer spending.

It is estimated that 92 million taxpayers will benefit from an average tax cut of US$1,083 each in 2003. Hopefully these tax cuts will encourage American consumers to buy more. But critics claim that the tax cuts will benefit the richer Americans who invest large sums on the stock market, whose dividends will enjoy tax benefits!

So the expected rise in consumer spending need not arise! *The tax cuts could actually be a political gimmick to win votes instead! But with the dramatic increase in military spending for the armed forces build-up to a possible war against Iraq, the U.S. budget deficit will deteriorate further.*

Already the U.S. budget deficit is estimated to slide to US$300 billion this fiscal year. How much more can this U.S. deficit be sustained without causing a loss of confidence and outflow of capital?

The European economy cannot follow the extent of U.S. deficit financing as it is tied to the E.U.'s Stability and Growth Pact, which caps national budget deficits to 3.0 per cent of the GDP.

Germany which is the biggest European economy, has the largest budget deficit. However, it is now being allowed some more flexibility, because of the of its long dole queues and as its public sector faces the threat of massive strikes!

The basic principle in the E.U.'s economic management is that "sound budget policy is the basis for durable economic growth" and they are trying hard to adhere to this fundamental fiscal policy!

These adverse economic international developments are having an unfavourable impact on the Malaysian economy.

At an ASLI Conference held in January 2003 on the "Malaysian Strategic Outlook", some financial analysts from investment institutions, as usual took a short-term view of the economic outlook. Unfortunately, these financial analysts are mainly concerned with the equity investment outlook for stocks and shares in the next few months. They are only interested in attracting the investment of short-term market players. Hence there is a tendency to "talk up the market".

There is less inclination to take a longer term and more realistic assessment of the economy, (especially if it is fraught with uncertainty and structural weaknesses) as it would spoil their market investment sentiment! This will not be good for their business of investing in stocks and shares!

Hence the chief economist of Credit Suisse First Boston, P.K. Basu, took an optimistic view stating weakly that, "with consumers' consumption getting stronger and stable, I expect the country's economy to be strong".

But K&K Kenanga head of research Seow Choong Liang stated at the same panel discussion that "as consumer sentiment was still low," he expected the economy to rebound less slowly!

Hence these short-term financial analysts often cause more confusion about the overall economic outlook, and thus create unnecessary uncertainty in the medium-term economic prospects!

With deflation and the prospects of war in Iraq, threatening the world economy, it is imperative that we take compensating policy action to counter these adverse economic developments or be dragged down into deflation ourselves!

If we are constrained by the budget deficits and the narrowing balance of payments surpluses, to undertake significant stimulative fiscal measures, then we have to boldly change policies and sacrifice "sacred cows" in order to fight the prospects of slowdown and deflation.

But unfortunately, we are not doing as much as we should to break the vicious cycle of losing business confidence and resisting the damaging deflationary forces!

For instance, in my January 2003 *New Straits Times* monthly column, I stressed the need to cut down red tape, so that the civil service could accelerate rather than decelerate the pace of socioeconomic development!

Interestingly enough on the same date on January 5, 2003, at least three well-known *New Straits Times* columnists had similar themes!

Dr Munir Majid stated that "we have to be more business-like and legalistic" in our dealings with Singapore and that we cannot "rely on *kawan sama kawan* (friend with friend) which has been a big mistake of the past"!

This is true for our whole administration where we cannot be just *chin-chye* (complacent) in our overhaul management. We have to become even more professional!

Then, Tan Siok Choo asked in her Weekend Guest column in the *New Straits Times*, whether Malaysian policy planners could "inculcate a culture of maintenance in the public, for everything they build, operate and use"?

Why do we not give sufficient priority to maintenance? Is it because the government does not provide enough funds? Is it because our leaders prefer to allocate funds for new projects that they can ceremoniously launch? Or is it because the funds that are provided for maintenance are hijacked for other purposes?

Perhaps it is a combination of all these and other reasons. Nevertheless, it is time that the maintenance is given higher priority or else there will be greater wastage of public funds in the future.

Ahmad Talib in his *Pahit Manis* column in the *New Straits Times* also touched on this theme when he quoted a poem: "We cling to good old values, we hold them tight".

However, we do not seem to keep the good old values, but instead choose newfangled ideas that get the economy into difficulties. *What about having, inter alia, more balanced budgets, cost effectiveness, accountability and transparency?*

With so much concern for the declining management and implementation capacity of the government machinery—what is the response of the government?

Should not these concerns stir our leaders to take strong and urgent initiatives to radically improve the government's administrative machinery, before it is too late? *The longer the delay—the deeper the rot will set in!* Then we may get to the point of no return or face the dreadful need to become too radical in our treatment of the decadence that could characterise the economy in the future.

It would be desirable to take action early to arrest the decline in the civil service which Dato' Seri Abdullah Ahmad Badawi will have to inherit when he succeeds Dr Mahathir. However, civil service leaders will have to continue to do their best to serve the government and our people in its efforts to boost the Malaysian economy.

A Mid-Year Review of the Malaysian Economy

What is our assessment of the Malaysian economy now that we are in the middle of 2003?

The Malaysian economy has weathered the storm of SARS and the adverse impact of the U.S.-U.K. invasion of Iraq. It has come out of its struggles in relatively good shape.

The government's economic package introduced last month will help the economy to recover from SARS and Iraq War shocks, but the damage cannot be entirely rectified.

Hence the earlier estimate made by the Treasury in the
Budget Speech 2003 last September, for a growth of 6-6.5
per cent for 2003, may not be realised, as seen in mid-2003.

Bank Negara's estimate of 4.5 per cent for 2003 will also
be difficult to achieve. But no one could have predicted the
adverse effects of SARS nor the continuing depressive and
deflationary global impact that has been worsened by the
U.S.-U.K. invasion of Iraq!

Most economists now consider the global and Malaysian
economic outlook to be less favourable. Economic growth
for Malaysia for the whole of 2003 could be in the region of
3.5 to 4.5 per cent, but only if we work hard at countering
the adverse effects of the first half of the year.!

In the light of the poor external economic environment,
the alternative for Malaysia is to more actively utilise all its
own human and capital resources and skills to develop the
domestic economy, and to reduce its high dependence on
foreign demand. But this is not being adequately done by the
government.

*The Yang di-Pertuan Agong Tuanku Syed Sirajuddin in his
Birthday Speech to the nation on June 6, 2003, wisely advised
Malaysians to support the government in its efforts to ensure the
well-being of the people, irrespective of race, religion, sex and
place of origin.*

The King also called on his subjects to improve
productivity and to make the economy stronger and more
competitive.

But is His Majesty's noble call being taken seriously? I
sincerely doubt it!

*Deputy Prime Minister Abdullah Ahmad Badawi made an
important statement at the Indian Chamber of Commerce 75th
Anniversary Dinner on June 5, 2003, when he said that "The
Barisan Nasional adopts an open attitude. We do not want any*

community to be left behind, educationally, economically and socially".

Coming from the Deputy Prime Minister who will take over the reins of the government in about five months in October 2003, his statement is indeed very encouraging. The full implementation of this policy statement will surely enhance national income and national unity.

But for now there are grave doubts about the actual interpretation and implementation of these noble objectives!

Unfortunately, there are wide gaps in many policies and their implementation. These painful gaps lead to the growing credibility gaps between policy statements and the reality on the ground.

This sad situation arises due to the government's declining delivery system. This has been honestly acknowledged by the authorities but has not been boldly addressed as yet. Why not?

The Public Service Department's Confession

For the first time as I can recall, the second most senior civil servant, the Public Service Department (PSD) Director-General, Tan Sri Jamaluddin Ahmad Damanhuri, has come out openly to confess that "despite cutting-edge technology, mission statements and Client's Charters, the standard of the civil service still leaves much to be desired"!

His statement to the one-day 'meet-the-people' session in Kangar on July 7, 2003, was somewhat of an understatement. Nevertheless, it was a bold and honest speech for which he should be commended.

Tan Sri Jamaluddin pointed out that "of the 1313 complaints received between January and May 2003, about 52 per cent were on delays, in responding to public enquiries or in taking action"!

The question that the public will ask is, "So now that you see the problem, what have you done about it or what are you going to do about it?"

The PSD chief urged civil servants to be hands on and to adopt "walk about management". He also asked what has happened to the decision to appoint an officer in every department to speed up the solving of complaints?

All this talk is fine, but the bottom line remains, What are the benefits for the public? Good intentions are not good enough!

What are the authorities doing about inefficient and indifferent secretaries-general and heads of department who are lax about the performance and discipline among their staff? Should they not be disciplined first?

Are the authorities reluctant to take action against a minority of civil servants who are derelict in their duties because of political repercussions? That should not be the case. The government stands to lose by mollycoddling its civil servants. Then decisions will not be properly implemented and the government will lose public confidence and support—even if all the one million civil servants vote for the government of the day—and that's most unlikely especially today!

A Tribute to Raja Mohar

This is why it is important for all Malaysians to draw on the inspiration and integrity of our lofty leaders of the past who contributed so much to build the strong economic, financial and other vital institutions in our country

This is therefore the time when we must pause to also pay a great tribute to the exemplary life and career of the late Tun Raja Mohar, who is internationally acknowledged as an outstanding civil servant.

We all wish we had many more officials with his intense devotion to duty, to serve God, King and Country to the best of his ability. His extremely high standards of quality, responsibility, integrity, humility, diplomacy and charm, are hallmarks that all civil servants would do well to emulate.

I had the distinct privilege to serve under Raja Mohar's enlightened leadership in the civil service, and know that we will need to follow his striking leadership qualities—if Malaysia is to achieve its Vision 2020 on time!

Indeed the government should launch a series of public lectures to honour and learn from his great contributions to our nation.

The Malaysian economy is gaining strength. However, it could soon falter again, if we do not speedily address some of the aforementioned outstanding sensitive and structural economic issues. Hence Abdullah Ahmad Badawi has to act fast to review, revise and restructure the Malaysian economy speedily, as there are also serious external threats to the sustained expansion of the economy.

This has been made clear by the June 2003 meeting of the Group of Eight (G8) countries, which showed once again how preoccupied these rich and powerful countries are, with themselves. They are almost indifferent to the prospects of the majority poor developing countries.

Thus they would be even more cavalier to Malaysia since it is a more self-reliant advanced developing country that wants "fairer trade and not aid"! We also need to expand our service industries like Education.

Private Education

The future growth of the economy will depend on how fast we increase the service sector's 56-per-cent share of the economy in order to achieve industrial status.

The education industry could provide great potential as a foreign and revenue earner, but are we doing enough to promote private education in the country? Much more needs to be done and at a faster pace, if we are to compete with many of our neighbours in the education industry.

The establishment of four education promotion offices in Saudi Arabia, China, Indonesia and Vietnam is a good move. But we must have more attractive education products to export.

The twinning degree programmes conducted with well-established foreign universities are highly regarded by foreign and local students. Their degrees are internationally recognised. We should therefore continue to expand these twinning degree programmes rather than constrain them as is the recent trend!

If we want foreign countries to recognise our local degrees and diplomas, we must raise our domestic academic standards and become even more meritocratic. But are we seriously willing to do so—and how soon?

The proposal to amend the Private Higher Educational Institutions Act 1956 to exempt foreign students from taking the local-based compulsory subjects (such as Malaysian culture and history) is welcome. It will remove any disincentives for international students to study in Malaysia, as they would want to study subjects that can give them employment.

Similarly, the proposed liberalisation of the present cumbersome immigration procedures for foreign students will make them feel far more welcome to study in Malaysia.

The establishment of the Technology Investment Fund with RM500 million to start with is also praiseworthy. However, the Fund would be more likely to perform better if it was privatised and run along business lines with the support of academics.

Matching grants to meet training costs are also beneficial, but hopefully the selected sectors will include education, health and the social sciences to promote national unity and better race relations and enable Malaysia to develop centres of excellence in these fields.

Food production got a major boost with an additional RM1.0 billion for its Fund for Food. The question that arises is how these additional funds are to be utilised more effectively. Will the high allocations for food production raise the productivity and ensure sustainability in agricultural production? If the Fund for Food is not managed efficiently, much of these large allocations can become counter-productive, as agriculture is a sensitive and difficult area to finance.

The construction and housing industry has also rightly been given greater incentives through tax exemptions and subsidies that will benefit the poor under the scheme for Home Ownership for People (Hope). The government's National Housing Company (SPNB) is expected to build 150,000 affordable houses in five years. Will this be achieved?

But unfortunately no announcement was made that the present building techniques which are slow, labour intensive, costly and inferior in technology will be changed. The government could have taken this opportunity to introduce the modern Industrial Building Systems (IBS), that would be more capital intensive, more efficient, faster and with higher quality standards!

Hopefully, it is not too late for the government to make it conditional for the contractors who win government housing contracts to use the IBS. This would optimise the use of public funds as well as the benefits to low-income house-buyers.

Fourthly, enhancing the effectiveness of the delivery system is perhaps the most important strategy in the Stimulus Package. The government can provide the best policies and packages, but in the end its success depends on effective implementation. This is where the civil service, as well as the private sector, will have to be far more efficient in the delivery of services on the ground.

However, there were no major measures to improve the whole civil service, and no specific policies to force the private sector to be more efficient and more competitive!

The new measures to expedite the approval of building plans and Certificates of Fitness for Occupation (CFOs) was most welcome. But there is no assurance that the desired results will actually be achieved, as there have been similar well-meaning efforts in the past that have fizzled out!

The non-operational Client's Charter in government departments are a testimony to the gap between policy and practice. What is really essential to make these good policies work is to enforce stronger discipline and accountability. Those who delay or cannot deliver should be penalised if not sacked!

Unless the government takes tougher action, the delivery system will continue to deteriorate, whatever incentives are generously offered to the public service.

As for the private sector, a more competitive environment and the natural quest for profits, will drive businessmen to work harder for their very survival, in these difficult times.

The reduction of EPF contribution for employees from 11 per cent to 9 per cent is necessary at this stage to stimulate consumption, but the same argument could be used for employer contributions too! But employers' rate of contribution will remain!

Some trade unions might understandably be unhappy as EPF contributions are meant to protect workers' welfare on retirement.

Private sector workers could feel left out as they will not be entitled to half a month's bonus that will be given to government employees whose pensions and employment are also guaranteed.

Hopefully government workers will be grateful for not having to show higher productivity to deserve a bonus, when their counterparts in the private sector may suffer zero bonus, pay-cuts and even retrenchment!

Bank Negara's Special Relief Guarantee Facility of RM1.0 billion is a significant contribution towards helping the badly battered tourism sector. The scheme will have to be quickly delivered in order to maximise the benefits of the facility.

Bank Negara's cut in the Intervention Rate by 50 basis points will also provide stimulus to the economy. The Base Lending Rate (BLR) of commercial banks will now be reduced from 6.42 to 6.0 per cent.

But the deposit rates will be also reduced. Hence Banks will continue to be assured of high profit margins. This is not proper, considering how much more difficult it is for businessmen to make profits!

Hence Bank Negara will need to refine the balance between commercial bank profits and their obligations to promote business and economic growth.

Overall, however, the economic package was impressive. Nevertheless, the success of the package will depend on the effective implementation of these stimulative policies and measures. Only time will tell if the government and businessmen will be able to deliver the economy from its present malaise and decline?

*With the right values and a stronger sense of urgency,
Malaysia will be able to overcome its current difficulties.*

Who Benefits from Government Policies?

It is also important that all policies need to be assessed in
terms of who actually benefits from these government
policies. What proportion of the genuine poor gain from
subsidised loans, micro-financing, scholarships and
community projects. *Are these benefits distributed according to
basic needs or according to other considerations?*

Are the beneficiaries ethnically balanced? Are Budget
expenditures being fairly distributed to all Malaysians,
regardless of race as stipulated in the NEP? This is doubtful as
there is much public dissatisfaction with the allocation of
public funds.

Are we giving the proper priority to Budget
expenditures? If so, why do we have these embarrassing and
disruptive floods in our country's capital city Kuala Lumpur?

If the constant flooding is due to poor environmental
management, especially by the state governments, then why,
for example, is the federal government unsuccessful in
pulling up the states and the local authorities for all their
gross negligence and irresponsibility.

The public needs to see more transparency and
accountability in the utilisation of their taxes—if they are
not to lose confidence in the government's management of
the economy!

*But why is the public so timid in their protests of what may be
perceived as questionable government expenditure priorities and
policies?*

Is it because we are complacent? Then we deserve the
governance we get. Or is most of the public passive, for fear

that their criticism may suffer from repercussions under the
Internal Security Act 1960?

*It is far-fetched and a gross exaggeration to imagine that the
government will penalise its citizens for airing legitimate
grievances!*

Nevertheless, the appropriate authorities have to put our
economic house in order, before Malaysians get more
impatient and express their disenchantment through the
ballot box!

Unless we faithfully examine these public policy issues,
and become more transparent, it is likely that cynicism of
policy statements will rise and credibility in public policy
statements will deteriorate, to the detriment of public
confidence in good governance!

There are many who feel "left out" in the full
participation in the development of the economy.

Thus we may not be utilising the full potential of our
national human and capital resources and talents, for the
accelerated growth of our economy and the attainment of a
higher level of social justice. For instance, students with
even more than 9As in the SPM are not given scholarships.
Many outstanding students are not given places in our local
universities.

The classic complaint was made by the Malaysian Indian
Congress that only one Malaysian Indian student out of an
intake of 200 medical students was accepted into the
University of Malaya in 2003!

How does this happen? Are the top Indian students so
uncompetitive? If so the results and full details should be
made transparent, so that public doubts and fears are
removed. Only then will the policy of meritocracy gain
credibility! Equating the STPM with the matriculation
examination is not meritocracy!

Then again how much is being done to recover the RM2.0 billion loans from the 100,000 defaulters, to help other needy students? Why are we mollycoddling our students and graduates? Are we teaching them the wrong values of "easy come, easy go"? If so, we will be spoiling them and turning out soft leaders in the future!

Meritocracy: Not Transparent?

It is difficult to explain to all those bright and poor students, regardless of race, that they have not got their scholarships or admissions to local universities, because of their poorer performance. Students know better. They compare their results with their peers and they find that they have often lost out to less able and more well to do students. Then they question the "meritocracy system" and lose faith in it!

How do we get our bright and poor students who feel left out or marginalised to show their loyalty to the country. Even the National Service (NS) programme will not help much to promote national unity, if there is perceived injustice.

National Service may even accentuate the feelings of "alienation" as the trainees would be trained to be patriotic to our country and yet they would feel that they themselves have not been given a fair deal by the country of their birth!

The meritocracy system has to be fair and open and transparent, to enhance national unity!

If non-*Bumiputera* students feel that they are being marginalised, they will not be so enthusiastic about national unity, or even of a Malaysian identity. Instead they may want to seek every opportunity to migrate.

This will be a sad situation for Malaysia. We will then lose many of our best and brightest, for no real good reason.

Then we must settle for mediocrity and not meritocracy!

Slow Progress of SMIs

Similarly, competent entrepreneurs of small and medium industries (SMIs) which do not have the right connections often complain of not being able to get bank loans—soft or hard.

So they are forced to go to the loan sharks, who will tear their businesses apart and weaken the backbone and competitiveness of our local industries! *If these SMIs are not encouraged, nurtured and well looked after (like the big multinationals that enjoy all the protection and tax incentives) the Malaysian SMIs will become less competitive and fade away.* Then what will we have—just big foreign multinationals that will run our economy!

But it may be worthwhile to institute an objective study undertaken by independent consultants to honestly find out why the technical assistance and considerable government funding are not reaching the SMIs. Is it because the officials who are mostly Malay cannot adequately communicate with the bulk of the SMIs that are essentially Chinese and Indian?

Or are there many other reasons for the low assistance given to our SMIs? We should thus be aware of the facts before we plan and implement our policies to assist our SMIs!

Worsening Corruption

Another threat to the economy is the growing problem of corruption. For instance, corrupt officials in local authorities, like the Ampang Jaya Municipal Council and many other agencies, bog down the approval processes, unless the officials are rewarded, directly or indirectly.

Prime Minister Dr Mahathir has warned that local council officials who are derelict in their duties will be replaced. But this is not enough! It would be better if they are severely punished for abusing public confidence and the people's taxes.

Let's see how many local authorities and other officials will pay the price for their negligence and corrupt practices? Or will the anti-corruption campaign be a flash in the pan that will soon be forgotten?

If that happens, public confidence in government's seriousness to stamp out corruption and inefficiency will surely wane!

Because of the growing gap between policy and implementation, even sound policy initiatives are greeted with much public doubt these days! When the public doubt keeps growing, public confidence and faith in the government's capacity to manage the economy efficiently, will decline. This in turn can cause social instability and weaken the government.

Environment Award: Mixed Feelings

We have to fight the malaise that is slowly settling over the inefficiencies of the government machinery.

Thus the launch of the Bandar Lestari Environment Award for cities and urban centres by the Minister of Science and Technology and Environment Dato' Seri Law Hieng Ding, on World Environment Day, has been received with mixed feelings.

Is this another policy with good intentions that is doomed to failure due to poor implementation and weak sustained enforcement? Will all the fanfare at the launch of the new award be lost, with the fizzling out of the enthusiasm soon after the opening event?

The Minister himself admitted on World Environment Day that the "laws are there but enforcement is another issue"!

The Director-General of the Department of Environment (DOE) Rosnani Ibrahim also stated that "city authorities could frequently get comments from the public to better tackle environment problems." But the public is getting fed up of complaining because its difficult to get their comments across and when received they are often ignored! So the public is reluctant to complain and so becomes indifferent, antisocial and even anti-establishment!

The recent devastating flash floods in the Federal Capital is testimony of the low priority that is given to environmental protection!

Hence there is rising public disillusionment in the implementation capacity of the government, especially at the lower levels of administration, like the local authorities. Unfortunately, this sentiment sometimes translates into distrust of politicians and their promises and policies!

This trend can have serious implications on good governance and the political leadership at all levels. This declining public confidence must therefore be arrested before further deterioration sets in and it becomes too late to rectify the steady slide in public confidence!

To enhance public confidence in the government's will to protect the environment, there has to be much stronger enforcement. Grassroot support from NGOs like the highly committed Environmental Protection Society, Malaysia, must also be strengthened with reasonable government financial support. Public complaints and criticism should be actively encouraged to sincerely speak up without the fear of negative reaction!

The laudable aims of the Bandar Lestari Environment Award could fail unless the government shows a stronger will to implement its good policies more effectively, with wider support of the ordinary people on the ground.

We must act fast as any neglect of the environment will add to the cost of doing business and sharply erode our international competitiveness!

As Prime Minister Dr Mahathir pointed out at the Transparency and Integrity Dinner to honour three great leaders, Tun Dr Ismail Abdul Rahman, Tun Tan Siew Sin and Tun Ismail Mohd Ali, for their impeccable integrity, that corruption is a poison. But are we doing enough to prevent corruption from killing the economy softly and even undermining our national sovereignty? I believe much more needs to be done to strengthen our economic resilience, especially when the G8 countries are self-centred.

Failure of the G8 Meeting

The Group of Eight (G8) richest and most powerful countries in the world failed us all by issuing a lopsided Final Statement at the close of its Heads of Government meeting in Evian, France, on June 3, 2003.

The final statement was devoted almost exclusively to warning Iran and North Korea about undermining "agreements to prevent the spread of weapons of mass destruction". *It hardly mentioned any measures to combat poverty in the Third World.*

The G8 also resolved to intensify the fight against international terrorism, but did not decide on any policies to remove some of the root causes of international terrorism, one of which is poverty.

The high trade barriers and the stifling agricultural subsidies imposed by the rich countries against the poor Third World countries were not given any priority in the First World's self-centred discussions.

This indifference to the plight of even the world's poorest developing countries, will widen the many gaps in the basic needs and fundamental human rights of Third World citizens.

These socioeconomic gaps in income, food consumption, health, education, environmental quality and the general standards of living and human welfare will create more fertile ground for the spawning and nurturing of international terrorists.

Then no amount of money spent on wasteful armaments can win the war against international terrorism.

What the rich G8 countries need so urgently to do, is to open their markets to the poor countries and to withdraw their heavy food subsidies, to enable the poor countries to raise their standards of living and thus to reduce poverty, which is a major root cause of international terrorism.

Only when the G8 countries see the wisdom to bring about a more just and equitable world, will there be peace, stability and prosperity for the whole world to enjoy.

In the meantime, Malaysia is continuing to liberalise its investment ownership policies to expand economic growth and raise living standards.

100% Foreign Ownership Surprise
Minister of International Trade and Industry Rafidah Aziz announced on June 17 that foreign companies can now own 100 per cent equity in all manufacturing projects immediately.

This is a major breakthrough in our policy on FDIs, as this issue has been a bone of contention for foreign investors and a disincentive to attract more foreign investment into Malaysia.

The 100 per cent foreign ownership will hopefully draw more FDIs, especially at the higher end of the value chain, like photonics and wireless technology, because our labour costs are not as competitive as in China and Vietnam.

Thus domestic investors who are joint partners with the foreign investors will also benefit.

To encourage this trend, the Industrial Coordination Act should also be liberalised. As it is, SMIs with less than RM2.5 million capital will have to reserve 30 per cent of their equity for *Bumiputera* participation. This requirement compares unfavourably with the 100 per cent foreign ownership that has now been allowed.

Thus Malaysian SMIs will feel relatively alienated and lose their incentive to invest in Malaysia.

They may then take their investments to other AFTA countries and export their products back to Malaysia and other countries. Worse still our own investors may come back to Malaysia to invest as "foreign investors" if they take a minority share in a foreign joint enterprise that invests in Malaysia!

Hence our aim to build up our own domestic industries will suffer. We could therefore have foreign domination of our manufacturing sector, with the further repatriation of foreign exchange, that could undermine our balance of payments.

In the interests of equity and in order to encourage Malaysian SMIs, the ICA should also be liberalised, to treat our own investors as favourably as foreign investors.

But all this needs political will and UMNO's support.

Dr Mahathir's Final Speech at the 54th UMNO General Assembly

Prime Minister Dr Mahathir Mohamad made his last opening speech at the 54th United Malays National

Organisation's (UMNO) General Assembly on June 19, 2003, as UMNO's President.

Dr Mahathir will retire in October 2003 as President of UMNO and Prime Minister of Malaysia, and Abdullah Ahmad Badawi will take over both these vital leadership posts.

Thus it is important to highlight some of the major issues that Dr Mahathir raised in his last opening speech as it is most likely to provide the basis for the next Prime Minister's and UMNO's future thinking, at least in the short to medium term.

Dr Mahathir obviously wanted to leave his intellectual legacy for his successors to build upon. *Some of Dr Mahathir's significant thoughts about the present world order and the challenges that we will face, are outlined as follows:*

1. Although Malaysia and the world's economic structure has changed dramatically since our Independence in 1957, the challenges to our survival as a nation still loom large.

2. Europe, including Britain and those countries that were colonised by Europeans like America (including Canada), Australia and New Zealand (or the White People), will want to dominate and exploit the rest of the world, especially the Southern developing countries, including the Arab countries!

    European culture is based on Greek and Roman empire values which were grounded in warfare and conquest and colonisation, to exploit captive resources, to perpetuate their dominance.

3. Thus Malaysia and other developing countries will have to understand the Western mind and its aims to recolonise us directly (through wars as in the

recent developments in Afghanistan and Iraq) or indirectly (through globalisation, the World Bank, the IMF or the WTO and other international organisations which the Western world dominates).

4. *Domestic challenges too abound in many forms—extremism, racialism and disunity. I might add corruption, human-rights abuses, unequal income distribution, discrimination of different kinds, inadequate competitiveness and insufficient meritocracy.* These challenges will have to be overcome, otherwise our national resilience will weaken in the face of the external challenges.

5. Although Malaysia is a Third World country, many Malaysians often expect First World infrastructure and facilities.

Thus the pressure on us to live beyond its means will increase unless the people are made to realise that good governance dictates that we have to follow the sound maxim which is to cut our coat according to the cloth available. "The Malays and other *Bumiputeras* have become too dependent and do business only with government. Without the public sector and the NEP they will collapse. This threatens the future of the Malays, the *Bumiputeras*, Malaysians and our beloved country," said Dr Mahathir!

This is a very bold and honest statement which perhaps only Dr Mahathir can openly state without fear or favour—and remain unchallenged!

However, it is one of the most important Malay issues that his successor Abdullah Ahmad Badawi will have to contend with, without causing undue reaction.

It is therefore imperative that UMNO and its leadership has the right spirit and the political will to face these challenges, otherwise UMNO will fade away slowly.

Dr Mahathir thus urged UMNO to be united and to grow from strength to strength with the support of its Barisan Nasional or coalition partners. The government will weaken if UMNO and the Barisan Nasional become disunited and lose the support of the people.

These major challenges that Dr Mahathir put forward to UMNO delegates have a message for all the Barisan partners too. The disunity within MCA, the MIC, Parti Gerakan and other coalition partners, could deepen and undermine Barisan Nasional and national unity.

The greatest tribute that can therefore be paid to Dr Mahathir by all his Barisan Party members, is to close ranks and work more strenuously to strengthen national unity.

This will be the most important challenge for his worthy successor Abdullah Ahmad Badawi. He must be supported by all Malaysians in his efforts to overcome the vital challenges in order to strengthen national unity through a new economic agenda!

CHAPTER 8

THE NEW PRIME MINISTER'S ECONOMIC AGENDA

WHAT could or should be the new economic agenda for the next Prime Minister Dato' Seri Abdullah Ahmad Badawi—from November 2003?

Dr Mahathir Mohamad retires in October 2003, after a distinguished career as Prime Minister of Malaysia for an unprecedented 22 years of steadfast and selfless service to the nation. He can easily be called the *Father of Modern Malaysia* (*Bapa Malaysia Moden*)!

But he has elected to retire and we have no choice but to reluctantly accept his decision with gratitude for his great services to our country and people.

So now it is time to move on with the making of history for our country.

Hence now is also the time for his successor Abdullah Ahmad Badawi to review our economic policies in the light of increasing globalisation and to finalise his New Economic Agenda for Malaysia. What should the new agenda be?

On the economic front, there is much to be done to consolidate and build up upon the status (that Dr Mahathir has built for Malaysia) as an advanced developing country. We need to follow his innovative leadership if we are to attain developed-nation status by 2020.

Is this ambitious goal achievable on target?

It all depends on the world economic outlook and the performance of our own domestic economy.

At the present time, the U.S. and other major industrial economies are still not recovering strongly enough.

The U.S. economy which is supposed to be the engine of world economic growth, is expected to expand by only about an average of just 2.6 per cent per annum in 2003.

But what is more worrying are rising U.S. unemployment and the widening budget and balance of payments deficits, which could undermine the U.S.'s economic recovery and prospects for sustained expansion!

Although the official unemployment is estimated at only about 2.4 per cent in 2003, Paul Krugman, the eminent U.S. economist claims that unemployment in the U.S., is the worst in 20 years because "people have given up looking for work"!

There is a real danger that the U.S. economy will weaken, particularly if the rest of the world begins to lose faith in the U.S. economy and the U.S. dollar. It could get worse if some important dollar earners like Japan and the petroleum exporting Arab countries, withdraw more investment funds from the U.S.! Disinvestment by foreign countries in U.S. Treasury Bills and U.S. Bonds can cause major damage to the U.S. economy.

Europe and Japan are still struggling to strengthen their economies, which are also characterised by low business and consumer confidence. This is largely due to the fear of international terrorism as is the case in the U.S., where the terrorist paranoia and phobia can be debilitating!

All these depressing international developments could dampen Malaysia's own economic outlook and restrain faster economic expansion in the longer term.

Indeed the better Malaysian economic performance currently expected in the short term for 2003 is largely due to the higher external demand for our electronic manufactures, the big bonus given to civil servants and the increase in our petroleum output!

However, investments expanded by only 0.4 per cent, and this is worrying! The wider trade surplus was achieved by an increase in exports by only 0.2 per cent, but a decline of 2.9 per cent in imports in the second quarter of 2003.

These are short-term contributors to economic growth and are not significant. They cannot sustain stronger and more resilient economic growth in the longer term!

We are still too preoccupied with the short-term and not sufficiently focused on the structural longer-term economic prospects.

The New Agenda

Thus, in order to strengthen the longer-term resilience of the Malaysian economy, a New Economic Agenda must be adopted by the new Prime Minister Abdullah Ahmad Badawi.

The New Agenda could include, inter alia, the following changes:

The first priority for Abdullah Ahmad Badawi would be to step up the implementation of the sound policy of meritocracy that has been boldly initiated by Dr Mahathir Mohamad but only recently.

The emphasis should now be on *quality* and not *quantity*—both in the public and private sectors.

1. **As a former civil servant himself, Abdullah Ahmad Badawi not only understands the civil service as an insider, but more importantly, he appreciates what must be done to ensure that the civil service actually delivers.** He also knows how civil servants can just say "Yes Minister!" and on the contrary, do what they really want to do!

    Therefore unless he revitalises the civil service and pulls up some of its powerful personal as a matter of high priority, he will find his policies and responsibilities as the new prime minister extremely difficult to fulfil.

    The public and the businessmen expect him to put the national house in order, so that Malaysia can really achieve the slogan *"Malaysia Boleh"*!'

2. **The Federal Budget can be a powerful economic tool to stimulate and sustain high economic growth and equitable income distribution.**

    But the Budget is weak since it has been in deficit for about five years! The challenge of the Budget 2004, that is to be unveiled as early as on September 12, 2003, is to reduce this rising deficit!

    For 2003 itself there will be a supplementary Budget of a large sum of RM4.5 billion—RM3.0 billion for additional development expenditure and another RM1.5 billion for operating expenditure! Where then is the budget deficit going to be reduced? And how is the increasing budget deficit going to be financed? Not by more taxes but by more borrowing! Thus public debt will rise!

3. **Starting from Budget 2004 therefore, there should be more priority given to privatisation.** This is necessary in order to take the strain off the Budget and to improve the cost effectiveness of public spending, and to compensate for a weak civil service and its poor implementation capacity. But privatisation policies and procedures have to be revised to ensure that the public get a better deal, in terms of price and quality of the privatised services. Drinking water and a whole range of other services, such as health, education, transport and housing can be privatised to increase efficiency in the delivery system.

    Since the rates and charges for privatised services will inevitably go up, the poor should be subsidised but not those who can afford to pay. The profits earned by the privatised projects should also be carefully scrutinised to ensure that there is reasonable profit and not profiteering, through exorbitant pricing!

4. **Hopefully the Budget 2004 will also change the priorities in spending.** The development expenditures could be reduced, while more funds could be provided to the Operating Budget to improve the quality of the social services to the public and to improve the basic needs or human rights of the mass of the people!

5. **Furthermore, the Budget should aim to give the public much higher value for money.** It is such a pity that after all these years, the Public Works Department (PWD) is only now insisting on qualitative standards for contractors to ensure that public spending is more cost effective! The public

will react adversely to more financial mismanagement, corruption and poor performance by dubious contractors, like the ones who built shoddy school computer laboratories!

The excessive tender bids of between RM24 billion and RM42 billion, submitted by only three consortiums for the North-South electrified Double-Tracking Railway Project is another case in point! What makes these tenderers think that they can get away with such high tender quotes? There could be "rings" formed by tenderers to have an understanding among themselves to overcharge the government. They could thus quote very high prices, so that they can all share the excessive profits when one of them wins the contract! That is why the government must have honest civil servants who will take the trouble to "see through" these unscrupulous tenderers, and call off their bluff or deceit!

6. **Again, despite so much public complains, we still have ineffective enforcement of many government policies, rules and regulations and considerable indifference to the implementation of the Client's Charter.** In fact, the Charter has now become some sort of a joke in some departments! More electronic government could help to overcome the indiscipline and lack of commitment of many civil servants who somehow hold the belief that the world owes them a living!

7. **Foreign Direct Investment (FDI) would need to be attracted back to Malaysia after being drawn to China.** The more liberal foreign investment guidelines, including 100 per cent foreign

ownership, and the removal of the need to top up the 30-per-cent *Bumiputera* equity share, will help to raise the confidence of investors. But implementation has to be sincere.

8. **Moreover, we would need to reduce our public rhetoric and soften our criticism against the West, however justified we may be.** Foreign investors and foreign governments can be oversensitive as they are not used to criticism from Third World countries!

9. **Security and public safety have also deteriorated badly.** Officials have explained that these deficiencies in good governance are due to the lack of funds and staff. So why not the government spend more on recruiting more staff. But the government select better qualified and more multiracial staff, to reduce the growing culture of "scratching each other's back"!

With the high graduate unemployment now, it is surprising to hear that it is still difficult to attract multiracial staff into the civil service unless, of course, the recruitment system is skewed or the promotion prospects for non-*Bumiputeras* are bad?

The recruitment of Chinese in the Police force is a case in point. Talib Jamal, the Federal Police Management Director, stated in August 2003 at the Unicop College in Batu Caves, "I do not know what else to do in order to attract these [Chinese] people to join the force"! Perhaps the police and other agencies should undertake a survey to find out why it is "difficult" to recruit especially the Chinese to join the police force?

My own understanding is that the non-*Bumiputeras* perceive that they are not wanted by the *Bumiputeras* in the public service. They have this perception that the *Bumiputeras* feel that the government or the public sector is reserved for the Malays and that the non-Malays can or should find jobs in the private sector. Thus many *Bumiputeras* will do all they can to discourage the non-*Bumiputeras* from joining the public service.

Even when, some non-*Bumiputeras* get into the public service, they find that they are often kept out of the mainstream of policy discussions and of career advancements!

The head of the civil service will have to ensure that all civil servants are fairly treated in recruitment and that they are promoted on real merit, before we can see positive response from the Chinese and others to get into the civil service.

The same arguments would apply to the larger recruitment and the retention of Indians in the civil service, but to a lesser extent as compared to the Chinese. But the problems could fortunately be even less in the case of the Indians! After all, the Indians are a much smaller and poorer community than the Chinese. They therefore pose no threat to the Malays and the *Bumiputeras*.

However, the need for the Indians to get into the civil service is much greater than the Malays. The Chinese can find jobs in the essentially Chinese business sector. Similarly, the Malays can gain employment preferentially from the public sector.

Now where can the Indians get their jobs? If the private sector Chinese employers are reluctant to give the Indians an even chance for employment, then it must be the government that has to provide the opportunities and the balance in employment!

10. **Greater priority should also be given in the Budget and beyond, to combat pollution and to enhance the quality of our environment and living conditions.**

Contractors and developers who strike up questionable arrangements with the authorities to run roughshod over environmental objections should be monitored and brought to book, regardless of their political positions.

What is the use of achieving high economic growth rates while our environment is neglected and actually allowed to deteriorate? What good is economic growth when we do not have the health to enjoy it?

The government has to seriously address these longstanding issues instead of sweeping them under the carpet—if the government's credibility is to be enhanced!

Finally, Budget 2004 and Abdullah Ahmad Badawi's new policies in the proposed New Economic Agenda, will have to consolidate the impressive economic gains made under Dr Mahathir's strong stewardship and continue to go forward.

Raising Patriotism

Budget 2004 will also have to lead the Malaysian economy onto higher levels of more sustainable, balanced and cleaner socioeconomic development, that will benefit the whole Malaysian society. Then national unity and also Patriotism will increase!

As it is, there are many Malaysians who do not feel a strong sense of patriotism because they feel marginalised from the mainstream of progress in the country. They either perceive themselves as not getting enough subsidies or being marginalised by government neglect.

Both groups have to be satisfied by fair and reasonable policies and practices. This is the challenge of our government's socioeconomic and political policies.

The government has to reform the education and employment policies to ensure that all Malaysians enjoy a sense of fair treatment in educational institutions and the job market. Even perceived feelings of unfair discrimination will lead to feelings of alienation that will erode the spirit of patriotism. That is one reason why there is a brain drain!

At the same time, it is vital that our new Prime Minister Abdullah Ahmad Badawi takes the initiative to improve relations with the Singapore government. If there is stronger goodwill on both sides there will be no need for Malaysia to have to build a "crooked bridge"! Indeed this would be testimony of the bad relations that existed between the two close neighbours, during the time of Dr Mahathir's and the Lee Kuan Yew-Goh Chok Tong period of leadership!

The Crooked Bridge Over Troubled Waters

Prime Minister Mahathir Mohamad stated on August 1, 2003, that "we are forced to build a crooked bridge", to replace the existing causeway. This old causeway which was built by the British before World War II, crosses the troubled waters separating Malaysia from Singapore.

Indeed this odd situation of having to build a "crooked bridge" comes on the heels of the serious disagreement and the crooked logic over the negotiations to urge Singapore to raise the price of water at 3 Malaysian cents per 1,000 gallons, that Johor supplies Singapore, for the survival of Singapore!

Apparently Singapore's Prime Minister Goh Chok Tong has refused to allow Singapore to participate with Malaysia to build the 31.5-kilometre straight bridge because he prefers to keep the causeway the way it is for "sentimental reasons"! So much for pragmatic Singapore!

But Malaysia wants to improve the communications between the two countries and especially to raise the standards of the Customs, Immigration and Quarantine (CIQ) complex at the Johor end of the causeway.

Malaysia also wants to enhance the environmental quality of the Straits of Johor and its marine life by enabling the sea water to flow right through the straits, instead of being blocked by the present causeway. This causes a serious pollution problem!

There may be other reasons for Singapore's objections, but these have not been made publicly known.

Perhaps Singapore is not happy that the new bridge would be built high enough for 25-metre-tall boats to navigate under the proposed new bridge. Thus they could transport cargo between the two Johor ports of Pasir Gudang and Tanjong Pelepas (PTP).

Singapore may regard this development as a further threat to the competitiveness of the Singapore Port?

However, in the absence of solid arguments by Singapore for the rejection or non-cooperation over this bridge proposal, it would appear that Singapore is being stiff-necked, stubborn and unduly difficult with Malaysia once again!

Malaysian Prime Minister Dr Mahathir, however, is determined to build the new bridge. He has said that "I think we do not want to involve Singapore in this [bridge] project".

The Malaysian Prime Minister has made this major decision only about three months before he retires at the end of October 2003!

He is so convinced of the need to build this new bridge, costing about RM1.0 billion, that he says "I hope to cross this bridge ... if I am still alive then", when it is expected to be completed by December 31, 2005!

Finally Dr Mahathir also joked, "I'll be more than 80 at that time, so I may need a *tongkat* [walking stick]" and that "if I can't use the *tongkat* to help me walk on the bridge, I'll use the *tongkat* on the people who built the bridge"!

I hope that well before that time the Singapore government will agree to cooperate to build the new bridge, so that there will be a "straight bridge", and that Dr Mahathir will fulfil his dream to cross it—without a *tongkat*. Then no one will need to feel the *tongkat* on their pate and both Malaysians and Singaporeans, will all be able to enjoy using a beautiful modern bridge and a new Customs Complex!

The ruggered relations between Malaysia and Singapore are not confined to issues within the two countries. Our different stances in our foreign relations are also to blame for the thorns in our sides!

Thus the diametrically opposite views in our WTO trade negotiations, add to the difficulties in our stormy relations.

This is clear in our resistance to the so-called "Singapore Issues" that divide the rich and powerful industrial countries, from the Third World countries. Singapore by all accounts is a developed country and it appears to toe the line of the West or the North.

By contrast Malaysia as a developing country, albeit an advanced developing country, champions the cause of the poor developing countries. Hence we find another situation of inherent conflict between Malaysia and Singapore!

Singapore Issues Will Undermine Developing Countries at Cancun, Mexico

Malaysia will reject the start of negotiations on the so-called Singapore Issues at the 5th World Trade Organisation (WTO) Ministerial meeting at Cancun, Mexico, in September 2003.

Minister of International Trade and Industry Rafidah Aziz pointed out at the UNDP Conference in Kuala Lumpur on August 7, 2003, that we cannot have negotiations when even the preparatory work for the negotiations have not been completed.

She is absolutely right. We can only hope that other Third World countries will adopt the same stand. They should counter the enormous pressures from the rich and powerful First World countries, to push for faster negotiations on the contentious Singapore Issues—when the developing countries are not ready!

The controversial Singapore Issues are the following: (i) trade and investment; (ii) competition policy; (iii) transparency in procurement; and (iv) trade facilitation.

These issues generally aim to enable the rich countries to have full access to developing countries trade and investment, and government contracts, by removing all protective measures that presently help the poorer countries to develop on a sustainable basis.

WTO and Terrorism

Open competition with the rich advanced countries will disrupt and undermine the socioeconomic structures of the poor countries. Maybe this is what the rich and powerful countries are intent on achieving.

But they cannot and should not be allowed to advance their hegemonistic ways!

The First World industrial countries want to get their interests served first without consideration for the poorer Third World countries. This is a dangerous trend as the Third World countries will become increasingly marginalised to benefit of the rich and powerful industrialised countries.

This can nurture and encourage and indeed fan the flames of international terrorism, as the poor and the weak nations feel an acute sense of economic exploitation and deprivation and even persecution! With their backs pushed against the wall of poverty and socioeconomic misery, the poorest in the poor developing countries will have little choice but to resort to terrorism to put forward their case for international economic equity and the fair distribution of global wealth!

The rich industrial countries are still resisting the appeals by the poor countries to increase the market access for their industrial goods from the poor countries. Developing countries have been urging especially the richer, powerful countries to reduce their agricultural tariffs and vast subsidies that undermine the economies of the developing countries! But it has been to little avail!

So what is the use of pressurising the poorer countries to start negotiations on the Singapore Issues, which will widen the economic disparities and further aggravate the economic injustice against the poor countries and keep them suppressed.

The WTO Ministerial Meeting at Cancun should therefore focus on serving the cause of international economic justice for the poor countries or be prepared to face the consequences of more intensified international terrorism!

This view is also supported by the remarks made by the Deputy Prime Minister Abdullah Ahmad Badawi at the 17th Asia-Pacific Round Table held in Kuala Lumpur on August 7, 2003.

He said, "Terrorism could be fuelled by many things: poverty, militant beliefs, deviant teachings, but nothing inspired and fuelled terrorism as much as oppression and injustice at home or abroad"!

He added that, *"Arguably, the single most important factor driving international terrorism and providing it with recruits, is the Palestine issue. The invasion of Iraq has only aggravated matters further"*!

It is hoped that the Western leaders will take heed of the views of the leaders of the Third World which have been suffering at the hands of the leaders of the First World industrial countries. They have consistently ignored the pleas for a more just, equitable World Order, for all countries to benefit!

I believe that if the Western world continues to ignore the root causes of terrorism, it will be seen by history to have condoned, if not indirectly aided and abetted international terrorism!

From now on, it will be useful if the Third World countries—and those Western countries and individuals who are supportive of the Third World—could insist that the WTO and other international negotiations should take into account the impact on encouraging international terrorism, if the poor countries are unjustly treated in these WTO negotiations!

In considering the implications on terrorism, both the rich and the poor countries will be taking a realistic view of the real world and will be able to better manage, if not overcome international terrorism.

Greater international competition under the WTO can reduce corruption. This is possible as with the wider opening up to international trade, it is more likely that the international pressure and sheer competition, will make corruption more difficult to grow.

But until we reach a higher state of economic development, we have to live with less international competition at home and try to combat and contain corruption to the best of our ability.

That is why the next Prime Minister's call to step up the fight against corruption is so welcome.

Trade Unilateralism Breeds Terrorism

The SAID countries strongly opposed the growing Unilateralism in WTO trade negotiations and the rejection of the philosophy that "might is right"!

This is a strong initiative that has been adopted that other Third World countries could follow in the WTO negotiations (that will take place in Cancun, Mexico, in September 2003) and in all future international negotiations between the rich and the poor countries alike!

But it is likely that the developing countries will once again succumb to the pressures of the rich and powerful countries, to accept their self-centred goals to maximise the interests of the First World at the expense of Third World countries!

Why do the developing countries give in to the unilateralism and bullying tactics?

The simple truth is that almost all developing countries are at the mercy of aid from the industrial countries. Malaysia is the only or one of the very few developing countries that does not receive any aid from the industrial countries.

This is why Malaysia can afford to be independent of these powerful pressures and can even resist their intrigue and manipulations. And this is why Malaysia and its top leaders like Prime Minister Dr Mahathir, are disliked by many rich countries, especially the U.S.!

The WTO negotiations are also based on selective consultations with a few countries, behind closed doors in what has come to be known as the 'Green Room'. This process is not only undemocratic but coercive as the powerful industrial countries like the U.S. and the U.K., as well as Germany, France and even Japan gang up to literally bully and squeeze all the concessions they can from the poor developing countries.

Nevertheless, some Third World leaders like Brazil, Egypt, India and Malaysia, did put up a tough fight. However, in the end, they lost out to the might of the rich and powerful countries that practically twist the arms of the developing countries and use tactics of the old imperialists to "divide and rule"!

Now hopefully China which is a new member of the WTO will add its considerable weight to the struggle of the Third World countries against the economic exploitation of the industrial countries.

We must remember that the West will try hard to win and continue to strengthen its economic imperialism and their hegemony all over the Third World.

French Criticism on the Handling of Terrorism

On the broader international scene, the recent criticism made by the French Minister of Defence Michele Alliot-Marie, to the Center for Strategic and International Studies in Washington, was very honest and insightful!

Alliot-Marie mentioned that "while terrorism was a great threat, its causes must be addressed". She identified a major cause of terrorism as the sense of frustration in the face of injustice and poverty.

Indeed the Americans and their close followers the British and some other smaller staunch supporters with blind allegiance to the U.S., like the Australians and the Japanese, seem indifferent to the suffering and deprivation caused by U.S. policies in Palestine, Iraq, and in many developing countries.

Until and unless there is a more honest recognition as to what the root causes of terrorism are and the necessary international policies are put in place, to remove political oppression in Palestine and the eradication of injustice in international trade, the threat of international terrorism will continue to threaten world peace! The French Minister of Defence is therefore correct and courageous in her honest criticism of the U.S. and other Western countries!

Hopefully, the leaders of the rich and powerful countries will shed their imperialistic international ambitions. They should see their way to enlightened and civilised policies, to introduce a new world order that promotes peace and prosperity for all people and not only for the rich and the powerful in the North!

We only hope that there need not have to be another tragic September 11 to bring greater enlightenment to powerful Western world leaders, who will then perhaps see the truth and the imperative to act with a stronger will to

address the root causes of international terrorism, one of
which is surely poverty!

Combating Poverty to Fight Terrorism

The *New Straits Times* editorial of January 22, has boldly
stated that poverty and deprivation provides fertile ground
for the manifestation of violence and terror.

Indeed the rich and powerful industrial countries should
honestly accept this basic wisdom. Then they would be able
to effectively lead the fight against international terrorism,
by attacking not just terrorists but one of its major root
causes—poverty, among most of the Third World
community.

Instead they are misguidedly and unsuccessfully fighting
only the symptoms of terrorism. Actually, the rich countries
are inadvertently strengthening terrorism by defending the
present unjust world economic order that perpetuates
poverty and encourages terrorism.

Globalisation and the international trade and financial
systems have to be reformed with a new and more
compassionate world vision, to fight economic exploitation
and political oppression by the rich over the poor countries.

Further procrastination on the part of the powerful
industrial countries to give the developing countries a fairer
deal, will only provoke more international terrorism and
prolong it indefinitely.

Hence the wide intellectual gaps between the recent
World Social Forum in Mumbai, India, and the World
Economic Forum in Davos, Switzerland, have to be closed
soon if the world is to have a more secure, peaceful and
progressive future.

The Prime Minister has passed the First 100 Days of his
leadership with flying colours. He has embarked upon new

policies and set new directions for the economy. We now hope that the government will be able to deliver all that it has set to do. However, the Prime Minister can be assured that all right-thinking Malaysians are behind him to achieve his laudable goals for the economy.

New Strategy to Counter Terrorism

However, there is one new strategy that Third World countries can use to try to convince the First World to give the poor countries a fairer and more equitable deal in international trade.

The argument is that the continued exploitation of the Third World through the denial of markets to their agricultural goods and simple processed manufactures will lead to greater Third World poverty which could promote and nurture and intensify international terrorism!

Indeed I believe that the worsening poverty all over the Third World and especially the widening income gap between the Third World and the First World, will provide fertile ground for the growth and robust development of international terrorism!

This is one of the major causes of international terrorism that has to be firmly faced up to, otherwise the U.S. and its allies will be chasing shadows, in wasted efforts to combat international terrorism.

But will the Western world and others accept this challenge squarely or would they want to take the easier line of meeting force with force, where they think they can win the war against terrorism?

There may be the possibility that terrorism has to get worse before the West and many other countries realise that the root causes of terrorism have to be unearthed and treated, before the rot damages the whole tree of civilisation!

In the meantime, the U.S. continues to act naively and assume that it can take the world for a ride as typified by Condoleezza Rice's constant defence of the indefensible stand of the Bush administration on the war in Iraq and international terrorism!

All-Out War Against Corruption

The Deputy Prime Minister Abdullah Ahmad Badawi has come out strongly in our own war against corruption, at the Integrity Management Committee Convention held on August 11, 2003, in Kuching, Sarawak.

He was absolutely transparent when he gave some alarming figures that "from 1998-2002, about 1,340 people had been arrested for corrupt practices and 50 per cent of them were government officers"! He added that "steps should be taken immediately to fight corruption"!

The Deputy Prime Minister's bold remarks were followed up with similar sentiments expressed by the Chief Secretary to the Government Tan Sri Samsudin Osman. He categorically stated that the "Integration and Management Committee at the federal, ministerial, state and departmental levels, will help wipe out corruption in the public service"!

But several questions rise in the public mind as to what initiatives will be adopted, when they will be implemented, and how effective these measures will be, in wiping out corruption?

The onus is on the civil service leadership as the political leaders cannot do much, unless the civil service cooperates by reporting and then taking tough action against their own corrupt officials.

However, the civil service will also want to see more evidence of a stronger political will, to arrest and prosecute politicians who may themselves be guilty of corruption. *The government therefore should show the public that they really mean business in stamping corruption this time around! Otherwise, the public will treat these statements as good intentions, that are not necessarily backed up with action. The situation should not be rhetorical!* Then the fight against corruption will suffer a further set back and credibility will decline. The public will feel let down again and will become even more reluctant to support the government's efforts to combat corruption!

Thus we sincerely hope that the civil service, for the sake of its own reputation and the welfare of the nation, will take urgent action to impose stronger discipline among its own ranks to fight corruption more seriously!

Corruption can also be substantially reduced if unnecessary and outmoded rules and regulations, controls and constraints are removed.

Topping Up the 30% *Bumiputera* Share

Minister of International Trade and Industry Rafidah Aziz made a much-awaited announcement that listed companies will from now not be required to "top up" the 30-per-cent Bumiputera equity ownership, even after it is diluted by the sale of the 30-per-cent Bumiputera share by Bumiputera owners. In a way this was an inevitable decision. But why it took so long and after so much dissatisfaction with the former policy is not clear.

The non-*Bumiputera* businessmen and investors (including foreign investors) have all along considered it unfair to be asked to top up the 30-per-cent *Bumiputera* share. *Bumiputera* owners, who had obtained the shares through the Ministry of Trade and Industry at low prices, often sold these shares at much higher prices to make a quick buck!

By so doing the total *Bumiputera* share of the corporate cake was never able to get to the target of 30 per cent. Each time the target seemed to be achievable, the goalpost kept shifting as many *Bumiputera* businessmen sold their allocated shares and then asked for more shares to make more money!

As Rafidah rightly pointed out, non-Bumiputera businessmen became very "irritated" by the insistence for them to give more shares to Bumiputeras by topping up the diluted portion of the 30-per-cent equity target!

Obviously foreign and even domestic investors found this policy unfair and unreasonable and showed their disapproval in the only effective way they know. They began to pullout, reduce their investment or just stopped investing in Malaysia in preference to investing abroad!

But the 30-per-cent *Bumiputera* requirement will remain for all new investment. This time, however, the new investors will not suffer from having to apply to the Foreign Investment Committee (FIC), to exempt them from topping up the diminished *Bumiputera* equity shares!

This will still be regarded as a constraint to investors if they compare their investment opportunities in neighbouring and other countries, which do not have this kind of 30-per-cent reserve equity requirement.

With AFTA now in full swing and the impact of globalisation hitting harder on our protective policies, it is likely that there will have to be more changes in store to enable Malaysia to overcome the "competitiveness and cost of doing business in Malaysia"!

For instance, Malaysia has been resisting the U.S. pressure to start the negotiations on the U.S. proposed Trade Investment Framework Agreement (TIFA). But other

neighbouring countries, like Thailand, Brunei and
Singapore have already got going on the TIFA negotiations.
Now Malaysia is the last of these neighbouring countries to
start the negotiations. Hopefully, we would gain from the
experience that these countries have learned from their U.S.
negotiations. Thus we should be able to get a better deal.

I have felt that, although Malaysia has been relatively
slow to liberalise as compared to some of our neighbours, we
are nevertheless moving faster than our domestic
socioeconomic dictates.

*It will be increasingly difficult to sustain the protective
elements in the National Development Policy (NDP) if we are
going to liberalise externally!*

How do we reconcile the faster rate of liberalisation
externally when our internal national development policies
are kept relatively intact? There will be conflict in
implementation and incompatibility and a decline in our
capacity to cope with more and faster liberalisation and
globalisation, that is also discussed at the Langkawi
Dialogues.

The Langkawi International Dialogue in Swaziland

Our concerns on protecting the *Bumiputera* corporate share
in the economic cake are also of great significance to the
African countries. They are in a lower state of
socioeconomic development and have drawn much
inspiration from Malaysia's considerable success in achieving
our advanced developing economy status.

Hence the importance for the African countries to learn
from our development experiences.

Thus the Langkawi International Dialogue, that has now
evolved into the Southern Africa International Dialogue, is
very pertinent for the African countries' socioeconomic

development. Our NDP could help raise the standards of their indigenous peoples, after so much exploitation by their previous white colonisers.

The Langkawi International Dialogue, that was initiated by Prime Minister Dr Mahathir Mohamad, has now been established firmly in Africa. Its African version—the Southern Africa International Dialogue (SAID)—was held in Swaziland from August 13-17, 2003.

The goal of these Dialogues is the promotion and practice of the concept of smart partnership, which postulates a win-win relationship for all countries, both rich and poor alike.

This philosophy includes the principle of "prosper-thy-neighbour", as opposed to the past and present orthodox values of "beggar-thy-neighbour" policies, as practised mainly in the rich and powerful West. The Western countries apply their self-centred principles in their international relations, particularly to the poor and weak former colonised countries.

The final statement at the conclusion of the SAID, called the Ezulwini Statement, clearly asked for major reforms of the world trading system through the WTO. *SAID has urged the WTO to work towards a "just regime of global trade and to commit the WTO negotiations to the smart-partnership approach"!* The challenge is to get the rich countries to adopt Smart Parnerships and to reject imperialism!

U.S. Imperial Ambitions?

U.S. National Security Adviser Condoleezza Rice mentioned to the ZDF German Television on July 31, 2003, that "the U.S. has no imperial ambitions"!

Rice seems to believe that she can fool the Germans and indeed the whole world with that kind of bald statement, when in fact most of the American administration's conduct in recent times, and especially during the Bush regime, has been quite to the contrary!

Or it may be that she is playing her role as dictated by the Bush administration to which she was "appointed" and not elected! It is a pity that she has to follow the dictates of the President and thus give the impression that even a bright professor like her can run the risk of becoming an "Uncle Tom" (or rather "Aunty Tom")!

What are the issues that gave Rice away and can also call her bluff? In rejecting the charge by ZDF television station that the U.S. can be compared to the Roman Empire, she said that the U.S. is not a bully and that "it does not mean that the U.S. does not value its allies"!

But at the same time she referred to the American help to create institutions like NATO and the Marshall Plan and the rebuilding of Germany. She said: "We're now trying to do that, in a sense, in the Middle East, with Iraq and with the Palestinian state and with what we've done in Afghanistan"! She added that "And there again, it is the spread of values that will make us more secure"!

The American imposition of its own values on other countries therefore is not regarded by Rice as "imperialism"! She conveniently and dishonestly ignores the illegal and unilateral war and occupation of Iraq by the U.S. without the authority of the U.N. Yet she expects the world to believe that the U.S. is not bullying the world and blatantly turning a deaf ear to its many allies and friends!

Rice's comments were endorsed by the White House which happily released the transcript of her TV interview. No wonder the whole world, except its small number of supporters of its war against Iraq, resents the present Bush administration so much.

At the same time, al-Qaeda and international terrorism must be revelling in America's monumental folly in having invaded and occupied Iraq unilaterally without U.N. backing! We can only hope—and pray—that the U.S., as the great superpower, will learn some lessons in humility and become less arrogant and more wise, if it is to remain a secure and safe superpower.

There is no doubt that we are living in dangerous times. Malaysia must face unprecedented international challenges on both the economic and security fronts. Dr Mahathir Mohamad has been able to keep Malaysia on top of these challenges. Can we continue to do so?

Yes, I believe Dato' Seri Abdullah Ahmad Badawi, our new Prime Minister, will lead us to greater heights!

CHAPTER 9

TAKING OVER THE REINS
OF LEADERSHIP

MOVING into October 2003, we anticipate the historic transfer of power to Malaysia's 5th Prime Minister, Dato' Seri Abdullah Ahmad Badawi.

He will take over the leadership of Malaysia from Dr Mahathir Mohamad in what must be a remarkably smooth transition by any international standards.

The great credit for this impressive achievement must surely go to Prime Minister Dr Mahathir Mohamad who has left a great legacy of strong socioeconomic progress and political stability.

Malaysians generally are grateful to Dr Mahathir for his many contributions to modernising Malaysia.

The smooth transition of power alone gives a powerful signal to the world at large that all is well with Malaysia. The incorrigible sceptics have again been proved wrong.

The international business and investment community will be greatly encouraged by this seamless transfer of power.

It has removed business uncertainty and strengthened investment confidence.

We salute the immense contributions of Dr Mahathir Mohamad and hope he will continue to give us the benefit of his mature wisdom. We wish him a well-deserved retirement from active politics.

At the same time, we warmly welcome the new Prime Minister Dato' Seri Abdullah Ahmad Badawi who will have our strong support and confidence.

The Malaysian public will carefully watch the new Prime Minister's performance in managing the Malaysian political economy in the next 100 days!

Abdullah has to have his own agenda for socioeconomic development as we cannot carry on the management of the economy in the same vein. Times are changing rapidly with globalisation—and new challenges are constantly emerging!

The change in direction and priorities have to take place in the First 100 Days. The changes will need to be seen and felt, if the change in leadership is to be meaningful.

But Dr Mahathir has already made several policy changes in consultation with his present Deputy Prime Minister Abdullah Ahmad Badawi.

In his Budget 2004 speech, Dr Mahathir stressed that the thrust of the Budget will be to continue with policies and strategies, to stimulate and accelerate domestic economic activities, with greater participation of all Malaysians in economic growth.

However, unless there is a more even playing field for all Malaysians to participate more equitably in domestic business and investment, this vital policy thrust will be thwarted.

A major structural weakness in the economy has been revealed in that 90 per cent of our exports are goods produced by foreign companies in Malaysia!

What will happen to our economy if many of these foreign companies decide to pack up and leave for different reasons, including the desirability of moving to China? Thus our private sector must be more enterprising.

Budget 2004

The Budget Speech mentions the need for the private sector "to act immediately to make a quantum leap to become the nation's investor, producer and exporter". But it is not very clear as to how this worthy goal is to be achieved so soon.

In identifying "capable individuals and private companies to undertake these initiatives", we must ensure that only the most competitive Malaysian businessmen are selected. Otherwise, it will be difficult to achieve our own globalisation goals.

There is a definite policy shift in Budget 2004, to attract FDIs to the service sector, like banking, insurance, education, health, tourism, shipping, etc.

It is even more significant that the decision has finally been made to set up a "one-stop agency" for the service sector, to expedite project implementation, like the Malaysian Industrial Development Authority (MIDA) has done so successfully.

Here again it is important to staff the proposed new agency with the best qualified and experienced Malaysians, regardless of ethnicity, for it to really succeed from the very start.

The policy to emulate other countries where the domestic market has provided the base for local industries to venture into the global market, has to be approached with much caution.

Malaysian consumers should not be required to pay unduly higher prices for lower quality goods and services. Too much protection can result in unacceptable inefficiencies where consumers will not get fair value for their money.

There has to be a reasonable balance, particularly since we are a relatively small economy.

Even when excise duties are levied on foreign cars from January 1, 2004, the sale of domestically produced passenger cars will continue to decline, as long as the quality and prices of our cars are not competitive.

Malaysia's success in leading the Third World countries in rejecting the Western proposal in the WTO, to open our markets for procurement of government contracts, is welcome.

However, we have to ensure that the negotiated contracts for our local businessmen are also competitively priced. The new policy to insist on more "local content" can also add to costs, which could erode our competitiveness.

The key to our future competitiveness will be our ability to raise our low levels of Research and Development (R&D) "to develop our own patents and brands".

But this ambition has to be backed up with further improvements in our whole education system. *From our primary and secondary schools to colleges and universities, we have to stress more on meritocracy and competitiveness.*

We have no alternative. We have to compete with the best international academic institutions and graduates if we are to actually succeed in R&D and innovation. Our education system has to be world class, but how fast are we moving in that direction?

The provision of double tax deduction for two years to employers who hire unemployed graduates is better than just giving them allowances to undertake training.

Furthermore, it would be more productive to encourage more *Bumiputera* graduate employment, in lieu of reserving the 30-per-cent *Bumiputera* equity?

However, if the graduates were of good quality in the first instance, they would have been employed rapidly!

Hence, the tertiary institutions must be penalised if they keep churning out poor quality graduates! Why should the universities utilise public taxes inefficiently?

The setting up of private commercial wings in hospitals is a protracted decision that is encouraging. It will help to retain more medical specialists in government and also could even attract those who are now in the private sector, to come back to government hospitals!

But this scheme should be carefully monitored, lest the low-income patients are given less priority than the richer paying patients in the commercial wings!

The Fund for Food has already received RM2.3 billion earlier and now the Budget will allocate another RM1.0 billion for the New Village Micro-Credit Scheme.

All this priority to the agricultural sector is fine. But what are the outputs and benefits from these vast expenditures? The government has to explain how much our food production has risen and whether the cost of food has consequently declined?

The Budget has to give more details for the performance of these huge expenditures, otherwise taxpayers will ask how much wastage is incurred and to what extent the farmers are actually gaining from the taxpayers' funds?

The same questions can be legitimately raised for the huge allocation of RM3.3 billion for the agricultural sector? The government and the farmers have to be much more accountable to the tax-paying public!

The proposed listing of the Federal Land Development Authority (FELDA) is an innovative new strategy. It will considerably increase the *Bumiputera* equity share, which does not rise significantly, since so much is sold out for quick gains.

Now the vast FELDA wealth will be included under the corporate equity structure and the achievement of the 30-per-cent *Bumiputera* equity share of the corporate sector will be accelerated.

But safeguards must be imposed to make sure that the 30-per-cent *Bumiputera* equity share is protected and not constantly sold out and diminished! *Otherwise, we will never be able to achieve this important NEP target and it will continue to be a contentious problem that will inhibit our socioeconomic growth!*

The establishment of 50 Community Service Centres under the new National Social Policy is laudable. The aim is to curb social ills particularly among the youths.

However, the root causes of these severe social ills have to be carefully researched into and addressed seriously. Do the youths believe that they have equitable access to basic needs and opportunities? *Do they feel alienated, oppressed and despised?*

As a member of the National Unity Panel, I believe that unless we examine these fundamental social issues in depth to apply the correct solutions, there will be a great deal of wastage in public funds, and dangerous social trends will undermine the national unity and stability that we so sincerely seek!

Low- and medium-cost housing will be taken over by the government's Syarikat Perumahan Negara Berhad (SPNB). This is good news for both low-cost housebuyers and developers who have been campaigning for this policy for a long time. End financing will also rightly be provided to EPF contributors by the MBSB.

However, developers will have to pay a "contribution" in return for not having to build 30 per cent of low-cost houses, if they choose not to build these low-cost houses.

Depending on the size of this "contribution", most developers will opt to pay the contribution, rather than to have to tediously build low-cost houses, which the Budget Speech admits, is actually the responsibility of the government.

Furthermore, since the government will be building these low-cost houses, the state governments will not be able to avoid alienating 'suitable' land for low-cost housing. How could developers build and sell low-cost houses that are sited on unsuitable pieces of land that are usually far away from centres of employment?

Similarly, the government should consider purchasing the minimum 30-per-cent equity share of newly built houses that are required to be reserved for Bumiputera buyers. If this reserved quota is sold to *Bumiputera* buyers within a reasonable time after completion, then this requirement will be acceptable by developers.

Unfortunately, the reserved house quotas for *Bumiputeras* are generally not sold during the initial acceptable period of six months. In the meantime, developers have to bear the borrowing costs with the banks. It is far worse if the houses are sold only after the initial six months!

At the same time, non-*Bumiputera* buyers are not able to purchase these houses, as long as they are on the *Bumiputera* reserve list. Such dog-in-the-manger situation causes considerable resentment.

This policy breeds inefficiencies, loss of profits and worse still ferments frustration against this well-meaning policy which is aimed to give *Bumiputeras* more chances to own their own homes, even at cheaper prices.

Abandoned housing is currently found in 204 projects, with about 66,000 units valued at RM5.6 billion! Budget allocations have been made for the rehabilitation of half of these abandoned units (36,000) costing about RM3.4 billion! *But the question remains as to how and why these projects were abandoned?* Were the contracts given to the competent and financially capable contractors? What lessons have we learnt to prevent this kind of massive wastage of public funds and taxes? We cannot sustain this kind of financial and social wastage for too long!

Cleaning the environment has been given greater priority with large allocations of RM1.5 billion for a modern incinerator plant and another RM1.9 billion under several agencies to improve the environmental quality. Of this amount, RM680 million has been provided for flood mitigation schemes, particularly to reduce the embarrassing flash floods in the capital city of Kuala Lumpur!

But what is needed is also to enforce the regulations on the cutting of hills and the illegal dumping into our already "dead" rivers! It is a waste of public funds to spend so much on alleviating the damage to the environment, which could have been largely prevented in the first place. Better enforcement could have reduced pollution and saved money. Of course, the real beneficiaries would be the contractors, who ironically are some of the worst polluters!

NGOs have rightly been provided significant Budget allocations of about RM96 million, for AIDS, Women and Family development, health programmes and for the disabled. This is most welcome as it also reflects confidence in the capacity of NGOs to deliver good services. The government will however have to ensure that its limited resources are given to those NGOs with the highest social priorities and that these funds are properly administered. Any financial scandal can undermine public confidence in NGOs that do not live up to public expectations of prudent financial management.

The promotion of arts and culture has been boosted by an allocation of RM80 million. This is good and should be sustained. *However, it is also necessary to take a more liberal view in this promotion of the Arts.* It should open up new horizons for true artistic expression, rather than stifle this promotion of cultural development through ill-informed indefensible bureaucratic constraints! Local authorities, like the Kuala Lumpur City Hall, should be more broad-minded in the approval of licences to stage cultural performances! The commendable aim to promote cultural development will need to be enhanced.

National Service (NS) has been provided RM300 million for the training of about 100,000 youths. The aim is laudable as it is to nurture the spirit of cooperation, national unity, integration and a healthy lifestyle.

It is imperative, however, to get the right kind of trainers for this important exercise, if it is not to become a waste of time and of public funds. In the past we had Civil Defence training and that was not very useful. I hope we do much better this time for National Service!

It is also important that concurrently, we improve policies that will make all our youths genuinely feel that they have a strong vested interest in training hard to defend our country in times of internal and external emergency. *However, they can be imbued with a high sense of real patriotism only when they believe that they will all enjoy the same full and fair opportunities to develop their talents to the full, without feeling any sense of alienation!*

Finally, the economy is estimated to expand by 5.5-6 per cent in 2004. This stronger recovery is based on the assumption that the private sector will take the lead. Private investment is expected to increase rapidly by 9.9 per cent and private consumption by 7.7 per cent!

These are optimistic estimates, particularly when public-sector spending is scaled down to rightly reduce the many previous budget deficits.

But the Malaysian economy can perform better if the new Prime Minister Abdullah Ahmad Badawi removes the many constraints on the private sector that inhibit its faster expansion.

Budget 2004 has been described as an Election Budget as it has been generous to all sections of the society. But what is more important has been the subtle changes in many policies which the new PM Abdullah Ahmad Badawi will no doubt follow and hopefully further alter, as he develops his own agenda and direction for the Malaysian economy.

The 2004 Budget's favourable prospects could be seriously undermined if the international economy falters, due to the failure of the WTO Conference in Cancun, Mexico.

The WTO Conference at Cancun, Mexico

The WTO Conference at Cancun on September 10, 2003, was not expected to succeed because the rich and powerful countries had not kept faith with their promises to promote development in the Third World countries at the last WTO meeting in Doha.

The U.S. in particular has flouted the WTO rules. It has become more protectionist by imposing a large tariff on imported steel and introducing a Farm Bill that raised Agricultural subsidies by 67 per cent or about US$6.4 billion in 2003!

U.S. farmers will receive a total of US$19 billion in farm subsidies in 2003. This is about half the total world aid to Third World countries!

As Sherman Katz, the international trade scholar at the U.S. Centre for Strategic Studies, has stated, "I think the President is instinctively a free trader with a strong streak of doing what is politically advantageous [for the U.S.]"!

The U.S. and Europe both have ignored the developmental interests of the developing countries but still expect them to contribute to the success of the WTO Conference in Cancun!

The WTO is dominated by the developing countries at least in membership as they constitute almost three-quarters of the total members.

However, the powerful Western industrial countries, realising this factor, have countered with another strategy to "divide and rule". Thus the U.S. has been working very hard to finalise "bilateral trade deals" with individual countries, to circumvent the WTO.

The industrial countries have far greater negotiating resources, experience and skills and usually overwhelm and out wit the Third World countries in finalising unbalanced and unfair trade and investment agreements.

Later on, if and when the developing countries realise that they have been taken for a ride by the rich countries and want to ask for a fairer deal, the rich countries often insist on sticking to the letter of the word and not the spirit behind it. If the poorer countries show some resistance, then the rich countries would go to arbitration at the WTO or threaten trade sanctions!

So where is the justice in the so-called free trade in the WTO, which also favours the Big Powers that strive to dominate international trade!

The new Prime Minister Abdullah Ahmad Badawi will have to ensure that the Malaysian economy will be able to overcome these international trade problems and injustices, and succeed on a substantial. There has to be a new agenda and the new Prime Minister has to adopt a New Way Forward!

The New Way Forward

The new Prime Minister, Dato' Seri Abdullah Ahmad Badawi, has been in office only since November 1, 2003! He will take sometime to get used to the high office and to fit into his new role as leader of the nation. But he has had very good guidance and invaluable experience as the Deputy Prime Minister to the previous Prime Minister Tun Dr Mahathir Mohamad.

Thus Malaysians in general and the business community in particular, are confident that Abdullah Ahmad Badawi is therefore well prepared to take over the reins of power, in his stride. Most foreigners interested in Malaysia's impressive progress also back him.

They all believe that Abdullah's calm and collected professional style of leadership will benefit the country greatly.

However, Malaysians and foreigners will be watching him closely in the First 100 Days of his premiership.

He will need to consult with all levels of society, to learn about their aspirations and expectations, in order to plan his strategies. This will ensure effective policy formulation and implementation, that would bring peace and progress, to all Malaysians.

He is likely to continue with his illustrious predecessor's policies for some time, before he determines his own policies and directions to meet the new challenges and the rapidly changing times.

Dr Mahathir, in his last address to the monthly gathering of the Prime Minister's Department in October 2003, suggested significantly that Malaysians want more success for the country but that "it will be the responsibility of Abdullah", after Dr Mahathir has retired!

Revised Affirmative Action

One of Abdullah's major aims would be to introduce a revised affirmative action plan, whereby:

Firstly, Malaysia's affirmative-action policy should continue to apply to all those who genuinely deserve the government's assistance, regardless of race, to break out from their poverty and poor standards of living.

Secondly, all those who have benefited from an education, especially at the tertiary levels, should be urged to compete on their own merit without undue aid and assistance. This policy would apply mainly to the *Bumiputeras. The "dependency syndrome" should be slowly but steadily phased out for the Bumiputeras, in the national interest of achieving our Vision 2020 goals on target!*

We should also move away from the mere attainment of accelerated economic growth targets and seek to achieve greater human development and a better quality of life for all Malaysians, and not only the privileged of all races.

Indeed it is quite feasible for Abdullah to devise a New Economic Agenda for better socioeconomic development, given the strong foundations in infrastructure. The national institutions, that have been laid by all his illustrious predecessors, should be strengthened, as they have weakened of late.

But only the new Prime Minister Dato' Seri Abdullah Ahmad Badawi can come out with his New Agenda. This should come out soon—and definitely in the First 100 Days of his leadership!

As of now, Dr Mahathir's considerable achievements and policies are there for all to see and appreciate. But there are also many pitfalls that we must guard against, lest we falter and fall in the future. There are many public concerns about Malaysia's future.

Public Concerns

The main public concerns are broadly:
The reduced safety and security, the rising corruption, growing polarisation and extremism, the decline of standards in our education system, deepening protectionism and economic inefficiencies, the widening income disparity between and among the different racial groups, and the degradation of the environment.

All this and other public concerns add up to some uncertainty and lack of confidence as to our country's longer term socioeconomic prospects?

Abdullah would therefore need to feel the pulse of the people on the ground, the simple and ordinary Malaysians, to find out more about their basic needs, aspirations and concerns. Policy planning should be both "from up-down and bottom-up"!

Then Abdullah would be able to come out more confidently with definite policies and plans to address these and other serious concerns, in order to imbue a greater sense of public confidence in our future!

We Malaysians need to know early enough where we are heading to? As long as there are lingering doubts and uncertainties as to whether our weaknesses will be put right, it will be difficult to build longer-term national confidence. *Any delay in building longer-term public confidence will drive away domestic investment and also not attract foreign direct investment. We have to constantly contend with external economic threats too!*

Jewish Boycott?

The latest threat of a boycott by the powerful Los Angeles-based U.S. Jewish lobby group, the Simon Wiesenthal Center, is an overreaction and most unfortunate, at it can discourage foreign investment!

The U.S. Senate's resolution against Prime Minister Mahathir's statement at the Organisation of Islamic Conference (OIC) in Kuala Lumpur that "the Jews rule the world through proxies" was also an unnecessary overkill on the part of the Jewish influenced U.S. Congress!

Fortunately, the Bush administration has resisted the Senate's proposal to withdraw the meagre military aid of US$1.2 million to Malaysia. *It would have been ridiculous if the Bush administration had given in to the Senate's naïve resolution against Malaysia. It would then have given credence to the widely held view that the Jews control U.S. foreign policy, particularly against Palestine and the Muslim countries and institutions, worldwide!*

But all this controversy in the U.S. does not help to increase investment in Malaysia! According to Dr Mahathir Mohamad, the Americans have, shown by their angry outbursts in Congress and elsewhere, that the Jews indeed have an unduly strong influence over U.S. foreign policies!

Perhaps Abdullah will have the same justifiable stand to take on behalf of the Islamic world.

However, I believe that he will express his criticisms differently and more quietly, to avoid upsetting the highly sensitive U.S. political leaders who may be under the financial influence of the Zionists and Israel?

Lack of Confidence in the Future

All the above economic and political developments cause some uncertainty for especially our bright young professionals and skilled personal who get caught up with a mood of declining confidence, in the future.

They will thus emigrate to greener pastures abroad where they see more secure and comfortable futures for themselves and especially for their children.

Already a large number of our best graduates have settled abroad. Many parents are also planning to send their children abroad for higher education with a view to encouraging them to settle overseas.

Furthermore, there are many prominent Malaysians who already have permanent resident (PR) status, especially in Australia, Britain and Canada, as they feel uncertainty for the future.

There has been a major brain drain over the years, which unfortunately is not openly discussed.

If these negative developments gather momentum, we will all lose out because the Malaysian economy will not be able to progress at the fast rate that we have achieved all these years since *Merdeka*!

It is therefore essential for Abdullah to come out clearly and resolutely with his new plans and strategies that will show the New Way Forward, that will inspire stronger confidence in the future!

All Malaysians must be made to feel that they all have a worthy place under the bright Malaysian sun. The Almighty has blessed us all Malaysians with more than enough to live in unity and harmony!

Malaysians of all races must genuinely feel convinced that the government will promote the welfare and progress of all its citizens and not mainly the Malays!

The growing perception that the government is primarily concerned with the well-being of the Malays, is not healthy and has to be addressed early. Otherwise national unity will continue to be jeopardised!

Abdullah as the new Prime Minister can ensure that the great trust (*amanah*) that the Almighty has graciously bestowed on him, will be discharged with justice, for All Malaysians regardless of race, so that they will all feel grateful that this great Trust is fulfilled.

We must all therefore rally around the new Prime Minister to help him fulfil this noble mission to serve Malaysia and all Malaysians with equity, justice and compassion, to achieve genuine national unity and continuing socioeconomic progress.

While most Malaysians bid a fond farewell and say a big Thank You to Dr Mahathir Mohamad for all his impressive contributions to our beloved country, we also warmly welcome Abdullah as our new Prime Minister.

At this crucial crossroads in our national history, we wish both Abdullah and Dr Mahathir, all the very best in their future leadership roles, at home and abroad.

May God also bless us all, during this critical transition period and beyond!

Our hopes for an even better future rise high as we heard the first public speech by the new Prime Minister Abdullah Ahmad Badawi in his home state, Penang.

The New PM's Maiden Speech

Our new Prime Minister Dato' Seri Abdullah Ahmad Badawi's maiden public speech, made in his proud home state Penang, on his first day in office as Prime Minister, was indeed gracious. It bore his hallmarks of simplicity, gratitude, humility and pragmatism. He said in his short 20 minute speech, "I owe my new success (becoming Prime Minister) to all of you! I will always remember that I am also thankful to Dr Mahathir for leaving a successful government".

Abdullah also invited Malaysians, with his now famous
words, to "Work with me and not for me"! His new call for
"*Maju Malaysia*" ("Progress Malaysia") and "*Berjaya
Malaysia*" ("Succeed Malaysia") could well be the new
slogans and battle cry for Malaysia in the new glorious era
under his premiership!

These are impressive historical statements which will go
a long way to set the scene for greater good will and ready
support from Malaysians of all walks of life, for his sincerity
of purpose and his noble national aspirations.

*However, Abdullah will need to improve and change some of
our policies and practices, to harness the full patriotism and loyalty
of all Malaysians, to enable all Malaysians, to march forward
together, with higher commitment and greater determination, to
further strengthen our national unity, peace and prosperity.*

Abdullah has made a good start. Let's all hope that he
will gather more momentum and public support, to lead
Malaysia forward to greater heights of achievements and
public well-being, as he takes over the reins of leadership on
November 1, 2003.

From Mahathir to Abdullah Ahmad Badawi

Our new Prime Minister Dato' Seri Abdullah Ahmad
Badawi has been in office for one month already.

Has there been any difference in policy or style? If so how
significant or positive have they been?

Indeed, there has been a change in the national mood.
People are happy that the change in leadership has gone on
so smoothly. Our new Prime Minister has taken on the
premiership in his stride.

His first public statements have been inspiring. They
have indicated the gradual shift to a New Agenda in three
areas, as indicated below:

Firstly, it was wonderful that the PM's first visit was to his home state Penang where he paid loving tribute to his mother. This act of filial piety touched all Malaysians.

His first public speech was made not to big business but to the humble farmers and fishermen, to whom he gave government flood relief grants. *This public gesture was important in that he showed that his heart is where it matters—with the people, especially the poor and the vulnerable groups!*

However, his maiden official speech was reserved appropriately for Parliament where he described his speech to Parliament and the people as "a symbol of my respect for this august institution"!

What was most striking was his solemn declaration in Parliament thus: "I am fully aware of the need for me to carry out my duties with integrity, trustworthiness, efficiency and fairness. I am aware that I will be assessed by the members of this House, by the people and above all, by the Almighty. *The position and the power that have been entrusted to me is a test from God for this humble servant.*"

Secondly, Abdullah told his ministers at his first Cabinet meeting that they should form task forces "to tighten procedures and reduce bureaucracy to fight corruption"!

In fact, these are some of Malaysia's main weaknesses—growing bureaucracy and rising corruption. *Indeed, the whole NEP will come to grief and the citizens will suffer most, if bureaucratic inefficiencies and corruption deteriorate further!*

Abdullah as a former senior civil servant understands and appreciates the serious consequences of socioeconomic instability and degradation that a weak and corrupt civil service can bring down upon our nation. The government's vital delivery system could collapse and cause chaos, if not anarchy, if the public service is not improved as a matter of priority!

We have not yet got to that bad state of affairs. However, Malaysia's economic strength and stability could steadily decline, if we do not arrest the declining social trends now!

Abdullah, in one of his first directives, has advised that the outmoded District Offices should be re-engineered. This measure is to be followed by reorganising the many inefficient local authorities.

But Abdullah has to seriously consider the replacement of appointed local government officials with elected councillors. Then they would be more accountable to the voters and could be thrown out of office by the voters, if they do not perform efficiently and honestly!

Alternatively Abdullah could dismiss corrupt and inefficient nominated municipal councillors, to clean up the system and to show that he means business!

The next step is to improve the generally lethargic counter services where the public have had a raw deal for so long.

His surprise visit to the Immigration Department was an impressive move!

It reminded me, as a former civil service colleague of Abdullah, of the late Prime Minister Tun Abdul Razak Hussein's lightning visits to government offices, even in remote *kampungs*, to ensure that development projects were properly implemented to directly benefit the poor and underprivileged!

However, at this advanced stage of our socioeconomic development, these highly publicised spot checks by our heavily burdened Prime Minister need not be necessary, if the civil servants are able to serve our King and Country, with greater dedication and diligence!

Why cannot the senior officers be held responsible for the poor services to the often harassed public and taxpayers! Do we need the Prime Minister to tell our senior civil servants how to run their departments?

Their promotions and bonuses should be more directly determined by their productivity and quality of service, to their pay masters—the taxpayers!

Simple common sense proposals made by the caring Prime Minister, like arranging separate counters for the handicapped, the aged, pregnant women and little children as well as senior citizens, are surely within the scope of civil servants! Or have they lost their sense of empathy, compassion and commitment to serve the people?

I hope all these new innovative initiatives will be introduced simultaneously and seriously, across the board in all government agencies that serve the public.

This can be achieved if the Chief Secretary to the Government Tan Sri Samsudin Osman is given full backing to hire and fire the minority of undisciplined and indifferent public servants. A worker must be worthy of his wage and rewarded according to his worth!

The civil servants have been given so many salary revisions and bonuses and other perks. But they seem to want more and more but do relatively less to improve their performance and productivity, in serving the country. This should not be tolerated!

Abdullah has proposed the welcome establishment of the Malaysian Institute of Public Ethics for both the public and private sectors to promote good governance.

It would therefore be useful, if Abdullah would encourage the public to provide more feedback to government, on corruption and inefficiencies in government services. Then the appropriate penalties and rewards can be widely applied, to enhance the efficiency of the whole public service.

The Public Complaints Bureau (PCB) should be able to provide Abdullah's office with enough evidence of dishonest and indifferent civil servants and their poor quality services to the public.

Indeed the large number of deaths on our roads could be reduced considerably if the government agencies exercised more dedication to duty and took more aggressive action against errant road users and road bullies.

Now that Abdullah has instructed that a conference be held to get the public views on how to reduce the tragic death and destruction on our roads, I hope the government will finally get tough in making our roads safe!

But as the Prime Minister pointed out, he has to be told the truth, so that he can make the right decisions. One way to get to the truth is to invite the private sector to give the PM frank feedback, on the standard of government services. He should be able to compare this feedback with government reports, which might not be so open and frank.

The worst kind of service he needs from his public servants is to get what he has himself called "apple polishers"!

Thirdly, it is also very salutary when the Prime Minister gave the assurance that "minority groups would not be sidelined".

If a poll were to be taken today, it will probably show that many among the minority groups in our country, feel marginalised by the way many policies are implemented, by over-zealous and narrow-minded politicians and civil servants. That is one unfortunate reason why some of our students abroad tend to unfortunately run down Malaysia.

We need to research into this "minority fatigue" in order to understand the truth in the state of affairs of our country in order to strengthen our national unity. Many Malaysians feel marginalised and therefore react against government policies and practices!

The Prime Minister has taken many other significant steps that reveal his new directions and style of management. He wants to reduce lavish government official openings and launches and the presentation of expensive souvenirs.

He asked the judiciary to change their image of being regarded by many as "not independent"!

He also urged cooperation from the authorities and requested them, not to provide "merely lip service".

But it is such a pity that some of our senior officials have to wait to be told to do what is right and proper, before they can move?

We have to be sorry for the new Prime Minister as he will soon become exhausted, if he has to look into the details of running the administration at all levels!

Abdullah's most important policy statement on economic planning however was made in the Dewan Negara when he tabled the Mid-Term Review of the 8th Malaysia Plan.

Bumiputera Equity

He said that Bumiputera industries were fundamentally weak and that is why Bumiputeras had not achieved the 30-per-cent equity ownership target. Thus the government will be setting up another unit trust called the Danaharta.

However, Abdullah will have to get honest advice that the shortfall in the *Bumiputera* equity target of 30 per cent is largely due to the quick disposal of their newly acquired equity by its owners, for easy profits!

If there was better management of these newly acquired Bumiputera shares, we could have achieved more progress than the present Bumiputera equity share of only RM73 billion or only 18.7 per cent of the total market equity.

The government now proposes to achieve the target of at least 30-per-cent *Bumiputera* equity by 2010. This is to be attained by providing at least 60 per cent of public-sector procurements to *Bumiputeras*.

In fact, some large government agencies are reported to be providing nearly 100 per cent of their contracts to *Bumiputera* contractors. However, many *Bumiputera* contractors who cannot perform gladly sell out their contracts to non-*Bumiputera* contractors who can perform for premium prices.

This is the *Ali Baba* practice that still persists, to the detriment of developing genuine *Bumiputera* equity ownership and skills. That is why we often have abandoned or badly built construction projects, which waste taxpayers' money!

The Prime Minister offered a noteworthy proposal to overcome this long standing *Bumiputera* problem, by encouraging the *Bumiputeras* to improve their skills and capabilities "with the help of the non-*Bumiputera* community which is further ahead."

This is excellent advice from the Prime Minister. So why not the government award the larger and more difficult contracts, on a priority basis, to Malaysian companies, with majority Bumiputera or government equity, so that all races can benefit from Malaysia's growth and prosperity?

This way, *Bumiputera* contractors can gain valuable experience and more equity ownership and the Chinese, Indian and other minority races would not feel alienated.

Some of the wastage of government funds could also be reduced, as the projects would be better managed.

Would Abdullah consider this win-win formula for sustained and ethnically balanced economic growth prosperity and national unity?

Malaysia's new PM has had a remarkably good first month of fair and firm leadership. The people are inspired by his positive and pragmatic speeches. However, the people will wait to see the actual follow-up and the clear evidence that the Prime Minister's sound advice is being efficiently executed, on a sustained basis, for the benefit of all Malaysians. *But already there are growing doubts as to whether the PM's new policies will be properly followed up and implemented on a sustained basis?*

Let's take a concrete case of the proposal for government to be a good paymaster.

The Government as a Good Paymaster

Prime Minister Abdullah Ahmad Badawi announced after the National Finance Council meeting on November 22 that the government will be a good paymaster.

On November 11, he instructed that the RM3.6 billion that the government owed to contractors from 2001 to October 31, 2003 should be settled by the end of 2003. He also stated that all government-appointed contractors would be paid before *Hari Raya*!

Thus, by November 20, RM2.4 billion was paid out and by November 22, the balance of the RM1.2 billion debt to contractors was mostly settled!

The Prime Minister and Minister of Finance has therefore delivered on his promise—even ahead of schedule. This quick response shows that the large and long overdue payments were really unnecessary—if there was a will to be efficient and to settle the payments quickly.

Hence it is unfortunate that the Prime Minister had again to intervene to get the government machinery to move faster, this time to settle its own debts!

What has gone wrong with the government's administrative system, particularly its financial management?

Delayed payments cannot be due to the lack of funds, as the government, with its strong economic fundamentals, has ample access to funding.

But perhaps the problem has been because of the weak administrative procedures in approving the release of the contractual payments.

Or are there some elements of deliberate delays in the processing and payment of bills, so that some unscrupulous civil servants can benefit from purposely delaying payments.

There also could be many contractors themselves who are guilty of poor workmanship, for which government payments should not be made as they would waste the taxpayers money.

Whatever the reasons, it is important that the government and/or the Auditor-General should carry out thorough investigations to find out the real causes for these unacceptable delayed payments, as soon as possible. The Auditor-General's investigation could also reveal outdated and cumbersome procedures that could be further streamlined.

This would increase the government's efficiency, transparency and also help to reduce bureaucracy and corruption.

Indeed, those officials who are found guilty of deliberate delays by not following contractual obligations to settle payments in 30 days, should be disciplined and penalised.

By being soft on civil servants who are derelict in their duties, the government itself runs the risk of being perceived by the public as condoning gross inefficiency and even corruption!

If these disciplinary measures are not taken early enough, the government will continue to be regarded as a poor paymaster. It will also not be seen as serious in practicing what it preaches! Then this kind of bad impression could undermine public confidence in the government's administrative efficiency. It could also adversely affect business competitiveness and the private sector's capacity, to be the expected engine of Malaysia's economic growth.

Hence this is a wonderful opportunity for the new Prime Minister to solve the perennial problem of delayed payments permanently!

But will late payments by the government persist in the future? Will new procedures be put in place to overcome the weaknesses that perpetuate late government payments?

Will the Prime Minister be able to put a stop to this embarrassingly bad practice that will erode confidence in government if the administrative weaknesses are not set right soon?

Will the new PM be able to pursue his new Agenda in his First 100 Days in office?

CHAPTER 10

THE PRIME MINISTER'S
FIRST 100 DAYS

PRIME MINISTER Dato' Seri Abdullah Ahmad Badawi
has completed two full months at the helm of the nation.
But the public is already beginning to ask how far he has
gone in delivering a new agenda?

His policy statements as Prime Minister have been
laudable. But have the government agencies begun to act on
his views and to seriously implement his policies at the
beginning of 2004?

There is not much improvement in the delivery of
government services to the citizens so far. Most people do
not see much change at the grassroots. But I suppose we need
to give more time to the civil service before it can respond
meaningfully.

However, the longer the delay in showing concrete
improvements in the policies and the implementation
thereof, the more difficult it is going to be to prevent doubts
in the government's capacity to deliver on its promises.

More delays in introducing reforms in the public service and in its outdated procedures and processes, will also make it more difficult to prevent the decline in the quality of service in many areas of the government machinery.

Fortunately, however, there are growing indications that Abdullah's government is continuing to introduce new concrete policies as part of his New Agenda, as evidenced below:

1. **Abdullah's biggest decision since he took office was to postpone the RM14.5 billion electrified double-tracking railway project.** This will warm the hearts of economists and concerned citizens who would like to see more priority given to reduce the continuing budget deficits.

   It is also heartening that the Prime Minister is closely reviewing the socioeconomic priorities in national economic planning.

   The huge resources that would have been provided for the large railway project can now be better utilised to achieve the more urgent basic needs of millions of our low-income and marginalised and vulnerable groups.

   There needs to be a full review of all the big projects that do not enhance the welfare of the citizens. Although the large cut back on major expenditures, will depress many big contractors, they should be satisfied with smaller projects that directly serve the welfare of the poorer segments of our population. They can venture abroad for the big projects!

   Big business must realise that they cannot prosper in the longer term, if the majority of our citizens, do not benefit more directly from huge development projects!

2. **Malaysia's decision to accede to the U.N. Anti-Corruption Convention is most welcome.** Now Malaysia's commitment to combat corruption will be subject to greater international scrutiny. Hence the government has to show definite results for all to see and to show the world that it means business.

   There was less obligation to do so before acceding to this convention, as the anti-corruption stance was not so clear and committed.

   The provision of a large sum of RM17 million to build an academy to train anti-corruption officials from Malaysia and abroad, will also show that the government has made not only a firm but a long-term commitment to stamp out the scourge of corruption.

   In keeping with this new proactive approach to combat corruption, it is hoped that our top leaders will also declare their assets and liabilities to an independent commission that will be directly accountable to the people.

   Then we can move with greater conviction and courage against all corrupt officials and individuals and companies, regardless of who they are, with a clearer conscience!

3. **Another important policy change is the automatic extension of professional visit passes for Buddhist, Hindu, Sikh and Christian priests, as well as to the substitute priests and musicians, serving at places of worship.**

The applications for first-time approvals and renewals of visas for priests can now be made at State Immigration Departments. This practice would considerably reduce the inconvenience currently caused by having to settle all these matters only at the Kuala Lumpur headquarters.

These innovative measures show pragmatism, empathy and religious understanding that is highly commendable. These moves not only help to strengthen religion and morality and national unity, but could also assist to improve our societal values and to reduce corruption.

4. **The decision to issue visit passes to foreign workers only after they have obtained a Certificate of Attendance of a two-week training programme will further smoothen the process of recruiting foreign workers.**

    The new policy will enable more workers to be recruited from more countries. We will thus reduce our dependence on just a few countries from which we employ our workers. Many of these workers have been involved in ethnic conflicts and criminal activities against Malaysians and we should reduce the intake from those countries which tend to send us problematic workers.

    Hopefully these new measures will reduce crime in our country and would give us greater safety in our homes and on our streets. This alone would increase public and business confidence.

    It is unfortunate that immigrant workers have been tolerated so long for their misdemeanours and their mockery of our national hospitality and generosity.

But why does Human Resources Minister Dato' Fong Chan Onn think that we need "at least a million foreign workers"? *Labour-intensive manufacturing and construction workers should be phased out and not increased.* How does this import of cheap labour help us to become more capital intensive and internationally competitive, through the greater use of science and technology, especially in the longer term?

Government will need to reconcile this serious dilemma of continuing to employ cheap imported labour, instead of adopting labour-saving technologies and high technological industries as soon as possible.

There should be new thinking and innovation in our production of goods and services that would need to be much more science and technologically based, if we are to raise our international competitiveness!

5. **The Prime Minister's call to hold a public forum to discuss the causes of the increase in traffic accidents and the tragic loss of lives, shows his deep commitment for public consultations and consensus.**

    Good governance implies that vital policy decisions are too important to be left too much to political leaders alone. The citizens have to live with the government's policy decisions, hence their views should be fully considered.

    One of the government's strategies to reduce road accidents and fatalities is to allow all the 82,000 policemen—and not only the 4,000 traffic policemen—to issue traffic summons.

This could be a very effective way of
encouraging road users to observe traffic rules.
However, it could also be a double-edged sword, as
there could be more corruption to contend with.
But the real goal is to reduce the tragic deaths on our
roads. So let's see if the death toll on the roads
actually declines substantially due to this new
policy?

Some cynics will call this a "win-win" measure,
i.e., we can have more corruption but less accidents.
Even if this is the case, we hope that the new policy
will work. After our traffic safety improves, we can
step up the battle against the corrupt givers and
takers on our roads!

6. **The Prime Minister's decision to establish a Royal
Commission to improve the public confidence in
the police force is laudable.** However, it will be
even more salutary if the government sets up a
Royal Commission to enhance the efficiency of the
whole public service, which largely determines the
future success of our country!

The whole public service is the backbone of
government and should be carefully looked after,
otherwise it will break the back of the government
itself!

7. **The Prime Minister is right in encouraging the
professionals in Bank Negara to determine if and
when the interest rates and the exchange rate of
the ringgit should be adjusted.** However, more
consultations between the government and Bank
Negara will help to develop a better perspective and
consensus on the need, quantum and timing of any
monetary adjustments. *After all, the steady weakening*

of the U.S. dollar, must have creeping adverse effects, on raising the cost of our imports and rising inflation.

The buoyant expansion of the Malaysian economy by an estimated 5 per cent and 6 per cent in 2004 and 2005 respectively, will also add pressure to the need to increase the interest rates and the exchange rates.

Hence we will need to be vigilant and timely in administrating monetary policy, particularly since the budget deficits are being attended to by the reduction in government expenditures.

The pressure on China to revalue the yuan is growing. There is a developing consensus that China may revalue its currency sometime just before the U.S. Presidential Elections in November 2004!

Hence Malaysia has to keep its options open and prepare to unpeg the ringgit to the U.S. dollar at the right time. At present, Bank Negara is reluctant to talk about adjusting the exchange rate. But the pressure will mount and a revision may come sooner rather than later!

8. **The continued generous tariff protection given to the national car Proton is a new challenge that Prime Minister Abdullah faces.** Proton has already asked for exemption of tariffs for another 20 years, since import tariffs have to be reduced by the beginning of 2005!

    Malaysia will lose credibility if Proton continues to be protected. The economy will have to pay the price for non-competitive production of the national car. *The better solution would be to form a strategic alliance with a major international car manufacturer and to bite the bullet of global competition?*

The imposition of excise duties on foreign or imported cars may be regarded unfavourably by the WTO. Excise duties are usually levied on domestic products and not on imported goods! The question arises as to whether we are postponing the day of reckoning in liberalising the motor car industry in Malaysia?

In the meantime, we are not exposing our national car to the sharp edge of full international competition. Hence when we have to really open our car industry to global competition, our national car could suffer even more, as we will not be sufficiently prepared to compete internationally.

9. **Another significant challenge is to respond effectively to the allegation made by the U.S. government that Malaysia favours certain religions.**

Thus the Prime Minister is right in inviting the U.S. authorities to visit Malaysia to find out the truth. If there were no religious tolerance in a multireligious and multiracial country such as ours, there would not be the national security and harmony that we enjoy.

But where there are lapses in religious tolerance, it would be useful and wise, to overcome the sensitive problems expeditiously. We have to counter this allegation by both word and deed.

At the same time President Bush and Prime Minister Blair better listen to the Bishop of Durham Tom Wright who described them as "a bunch of white vigilantes going into Brixton to stop drug dealing"!

The Prime Minister mentioned in his significant speech to the Christian Federation of Malaysia's

Christmas Open House: "Our multiracial, multireligious society is a gift we have been given to work with, not some kind of problem to be eradicated—and that in accepting this gift, we shall all be the better off for it!"

This is wonderful philosophy which when fully implemented will certainly make Malaysia a model nation of religious understanding in the world!

It is important to show the world that we have genuine religious understanding, to increase international confidence and to continue to attract goodwill and foreign investments.

10. **The Auditor-General's revelations at the 8th National Public Accounts Seminar have serious implications on the accountability of public funds.**

How is it that 39 ministries and federal government departments had not collected more than RM8 billion as of 2002? Why is this collection so bad?

Furthermore, the federal government has given out RM23 billion in loans to government agencies and the private sector interest-free or at low interest rates?

Are we deepening the subsidy mentality? If so we are then putting the nation in greater jeopardy and the new government has to take the necessary measures to reduce this subsidy syndrome. Otherwise we will be heading for financial trouble!

Government invested RM24 billion in 786 government agencies in 2001 and the audited reports of only 45 such agencies, from 1999 to 2001, showed that a number of them lost about RM2 billion!

This is a terrible waste of taxpayers' money. Should not some of these government agencies be sold out and privatised soon?

The government also provided grants from 2000 to 2002 totalling about RM5 billion to six companies under the Finance Minister Incorporated!

Are the managers of these companies capable of giving the public reasonable returns? Should we not go for meritocracy that the government has announced it will pursue? By rewarding a few so-called potential Bumiputera entrepreneurs, we could well be penalising the wider public interest!

In short, the question arises as to how long our economy can continue to prosper with this policy of high subsidies and heavy losses of the peoples' money?

And for how long can we sustain this drain on our national resources please?

I would appeal to the government to initiate thorough priority investigations of the financial health of these government agencies, to devise new policies to save the economy from severe financial strains in the near future.

I believe that the public will need a full explanation, so as to maintain public confidence in our financial management.

11. **It will be very beneficial if the government adopts this worthy practice of wide public consultations, on important issues that directly affect the welfare of the citizens.** This will enable the government to get stronger support for government policies which the citizens can identify with and support.

Prime Minister Dato' Seri Abdullah Ahmad Badawi has had a second successful month since he took over the leadership of the nation. However, we hope there will be more substantive changes in the near future. In the meantime, we wish our Prime Minister even more success in the future, as all Malaysians wish each other a better New Year in 2004!

The year 2004 will be the first new year with Dato' Seri Abdullah Ahmad Badawi as the new Prime Minister. We all hope that it will be the beginning of a new dynamic and progressive era for Malaysia—and all Malaysians.

But how has he fared in his First 100 Days?

How Much Has the Prime Minister Achieved?

The First 100 Days of Prime Minister Dato' Seri Abdullah Ahmad Badawi are over. How well has Malaysia fared?

Internationally, there is a warming up to his new style of reduced rhetoric, which gives foreign investors and some important foreign leaders more comfort and confidence in dealing with the new Prime Minister and Malaysia itself! Domestically, he has won strong public support and confidence for his new policy initiatives and firm leadership at the helm of state.

Policy Shifts

There have been at least three major policy shifts in his First 100 Days of premiership:

1. **There is clear evidence that his new agenda for the nation is to stress socioeconomic development that more directly benefits the human welfare of Malaysian citizens.**

Less emphasis is being given to building major projects as enough has already been achieved in the past. In the past, our infrastructure had to be built up to stimulate economic growth and to attract foreign investment. Now attention is being given to develop the soft side of economic development, i.e., human resources and social issues.

There is also welcome priority now being given to regard our multiethnic society as a God-given asset, rather than a liability. This fresh thinking could be developed into a Malaysian ethos. All our national policies and their implementation would also need to reflect this philosophy.

Thus the eradication of poverty regardless of race has to be given a new thrust in the fair and proper implementation of all our socioeconomic policies.

Similarly, the policy of meritocracy has to be effectively administered, especially at our schools and universities. This would be the best way to strengthen loyalty to the nation, and to enhance confidence in our future, for all young Malaysians.

Then there will be less talk of feelings of alienation, lack of confidence in the future and even emigration, and the consequent loss to the economy from the brain drain.

There is little point in providing attractive incentives to Malaysians who have settled abroad when we do relatively little to retain bright young Malaysians at home!

Our new Prime Minister will have to address these vital issues on national unity with greater priority.

In his earthy New Year Message, Abdullah said, "Let us stride into 2004 with a resolve to adopt noble values. A country whose people believe in God and respect the rule of law, will achieve greatness"!

These noble values are found in all our religions and cultures and enshrined in our Constitution, the *Rukunegara* and Vision 2020. These noble values are also found in the U.N. Declaration of Human Rights, which Suhakam has been established by Parliament to promote and protect.

Hence it is only right and proper that the new government gives much higher priority to attain a higher level of our social, economic and political rights in the future.

2. **There is clear evidence that the Prime Minister's new agenda for the nation is to stress socioeconomic development that more directly benefits the human welfare of the citizens.**

We know what these noble values are, but what we have not done is to give these noble values the necessary priority and the strong political will to implement them effectively.

Now that our Prime Minister has reinforced the national commitment to noble values, all good Malaysians need to give their full support, so that these values will be practiced in earnest and as part of our national culture.

However, the first step will be to ensure that the basic needs of the poor and deprived groups in our society are fully met and that they are not marginalised or left to feel alienated. If basic needs are neglected, our prospects for peace, stability and national unity, will be badly eroded!

Hence our socioeconomic planning and implementation have to be reviewed and redesigned, to focus on enhancing the welfare of the vulnerable and marginalised groups in our society.

We often forget that we are still a developing country and that there are about a million Malaysian households that earn less than RM1,000 per month. Given an average of about 5 members to a household, this will mean that about 5.0 million of our people belong to households that earn less than RM1,000 per month. This constitutes about 20 per cent or one-fifth of our population of nearly 25 million! This would mean that each individual in this category lives on an average of only RM200 a month, or RM6.66 or just over US$1.75 a day! That is about the income of half the world's population of poor people live on! Surely Malaysia can do better for its citizens!

We all know how very difficult it is for families to have a decent standard of living or a fair quality of life under these dire circumstances. This is especially so in the urban areas where you cannot even plant vegetables for your own needs!

We must therefore develop a new and effective Basic Needs Strategy, based on their Human Rights and the United Nations Millennium Goals, to deliver our poorest citizens from this dismal socioeconomic environment. We need to help them to get out of this vicious cycle of poverty and socioeconomic deprivation.

Prime Minister Abdullah Ahmad Badawi stressed in his remarks on the U.S. Bureau of Democracy, Human Rights and Labour 2003 that *"we plan development for all races, not just one race or religion"*! This is therefore the government's avowed policy which when fully implemented by the public service will go a long way towards strengthening national unity and *Bangsa Malaysia* (Malaysian Race), which we all aspire to achieve by the year 2020!

Like all those who attended the NEAC Business Dialogue with the Prime Minister in January 2004, I was very encouraged by the Prime Minister's frank and relaxed discussion on a broad range of issues raised by the corporate sector. However, there was major concern as to whether the sound initiatives taken by the new government would be effectively implemented by the public service.

Hence in order to avoid any creeping credibility gap, it is vital that the public service's implementation capacity should be considerably improved with a greater sense of urgency!

The Prime Minister's successful dialogue with the corporate sector should also be followed up with a similar dialogue with Malaysia's civil society. Datin Zaharah Alatas of the National Council of Women's Organisations (NCWO) has publicly called upon all women's organisations "to strengthen the principles he [PM] upholds, with viable action on our part"!

I hope our Prime Minister gives our civil society greater opportunities to better serve the underprivileged in our country.

One of the Prime Minister's most important decisions in his First 100 Days has been to appoint Dato' Seri Najib Abdul Razak as his worthy deputy and to reshuffle his Cabinet. From now on, the Prime Minister can depend on the Deputy Prime Minister to supervise the implementation of all his new policy initiatives. He can thus concentrate on the more important aspects of socioeconomic and political planning to meet the many new challenges that will emerge in our fast changing economy.

The change of Cabinet portfolios too will help to streamline the Cabinet and improve good governance and the overall performance of the economy. Thus the Prime Minister's call for his ministers to set the example for integrity, dedication and better delivery of government services to the citizens is praiseworthy.

Now the public service will have to serve the citizens with greater commitment. Too many politicians do not take their work to serve voters seriously enough. Too many of them are more interested in their own ambitions and the promotion of their own personal welfare.

However, to enable the public service to adjust more effectively to the new modern challenges of globalisation, it has become even more necessary to establish a Royal Commission or an independent committee to review the state of the public service and to make recommendations for its improvement.

3. **On the external front, the Prime Minister's official visits to Asean countries have earned a great deal of goodwill.** *Of special significance has been his visit to*

Singapore, where it was decided to raise the level of discussions to resolve our long outstanding differences.

This new understanding is most welcome. Now we hope that the old prickly issues will be removed promptly with pragmatism and statesmanship on both sides.

The PM's visit to Thailand was particularly important as it underlined a very important principle of promoting economic development: to address the issues of poverty and violence. Indeed the alleviation of deprivation and backwardness could also help to effectively counter terrorism in southern Thailand.

This strategy would be consistent with the call by the World Social Forum in Mumbai, India, held in January 2004, that called for "Another World Order", to replace the present New Economic Order that pushes for globalisation in the interests of Big Business, and ignores the welfare of the vast majority of the world's poor.

If only the rich and powerful industrial nations decide to fight world poverty through fair and equitable world trade policies, one of the major root causes of international terrorism will be removed! But, unfortunately, the rich and powerful countries are more intent on combating the symptoms of international terrorism, rather than the basic root causes of terrorism: political oppression and economic exploitation worldwide.

The political oppression is clearly seen in the Israeli Occupation and persecution of the Palestinians, with the liberal military aid and armaments provided by the U.S.

The economic exploitation is perpetuated by the WTO, which protects the agricultural subsidies of the rich countries in the E.U. and Japan and the U.S. On the other hand, they force the developing countries to liberalise and to open their markets for unfair competition from the rich and powerful countries.

In the near future the Malaysian government has some major challenges to deal with, such as the probable revaluation of the Chinese yuan and Singapore's many free-trade agreements.

China's Probable Yuan Revaluation

The U.S. pressure for China to revalue its yuan has been growing for some time now. But with the fast growth of China's economy at 8.6 per cent in 2003 and an estimated 8 per cent in 2004, inflation is likely to rise. This will break down domestic resistance in China to revalue the yuan.

But, in the meantime, the artificially depressed yuan has boosted China's exports. The U.S. balance of trade deficit with China has consequently been steadily rising from US$103 billion in 2003 to an estimated US$130 billion in 2004. At this rate the deficit will widen and within the next 12 months, both the U.S. and even China will find it difficult to allow the balance of payments deficits to grow further. Both sides will have to seek a compromise to protect their respective economies.

The war in Iraq and the huge expenditures on U.S. home security will have to be curbed to arrest the decline of the U.S. dollar.

Similarly, China will need to remove its fixed peg to the U.S. dollar so as to enable the yuan to rise and to float within a band of about 3-5 per cent on either side of the present rate of 8.3 yuan to the U.S. dollar, as in February 2004.

China and Japan which have the largest holdings of U.S. Treasury Bills could also reduce their investments in U.S. Treasuries and help to narrow the gap between the yuan and the U.S. dollar?

This currency adjustment could take place sooner rather than later and most likely before the U.S. elections in November 2004.

However, if China removes its peg to the U.S. dollar, other countries with similar fixed exchange rates would have to adopt more flexibility. Then the pressure to revise Malaysia's own fixed exchange rate will increase.

Hence it is better for us to be prepared for some departure from the present fixed peg of RM3.80 to the U.S. dollar, to that of a basket of currencies which would reflect the composition of our international trade.

But we could still maintain the necessary safeguards against irresponsible currency speculators, so that we can have a stable ringgit and be more structurally competitive in the longer term.

Malaysia has to closely monitor the regional and international currency developments as they could adversely affect our trade and economic performance in the future! But the forthcoming General Elections will do much to determine the new agenda for the Malaysian economy.

CHAPTER 11

THE GENERAL ELECTIONS AND THE NEW AGENDA

The General Elections were held on March 21, 2004, and we had just about two weeks to plan how to exercise our vote prudently. But in preparing to decide who to vote for, we had to ask ourselves, what and whether the new government is doing enough to deliver on its well-received promises to meet the aspirations of all Malaysians?

The choice of the "right" election candidates had a major impact on the votes that were cast for those politicians who can best improve our own and the national welfare?

The Prime Minister's sound criteria for the selection of the best candidates were therefore most timely and appropriate.

Malaysians diligently looked forward to the list of election candidates to ascertain if they are able, honest and dedicated to serving the best interests of all Malaysians.

The general good feeling was that Malaysia had achieved much in terms of infrastructure development. But there is a growing sense of foreboding, that we are losing out, in many areas of human welfare and the quality of our socioeconomic life.

The challenge for the new government then is to improve the balance between our rapid economic development and the necessary social justice, especially for our poor and marginalised groups, regardless of race.

There are many new threats to our sense of well-being. We can overcome them only through wider consultation with the NGOs and the people, to seek solutions where officials have not been very successful. We need new strategies to face these new threats to our socioeconomic progress and national unity.

Dato' Seri Abdullah Ahmad Badawi stated on January 31, 2004, just three months after he took over as Prime Minister, from Tun Mahathir Mohamad, "I have the duty now to deliver what I promised".

The Prime Minister's Agenda listed his promises as follows:

1. Changing the mindset of Malaysians;
2. Strengthening racial harmony;
3. Improving the public delivery system;
4. Enhancing transparency and good corporate governance;
5. Creating an efficient and "top class" civil service;
6. Eradicating public- and private-sector corruption;
7. Developing the country's human capital;
8. Providing more opportunities for Malaysians in the public and private sectors;
9. Reforming the education system; and
10. Developing agriculture as a key component of the economy.

It is too early and even unfair to make a full assessment of Abdullah's accomplishments. However, we all know that the feel good factor is strong and growing. That is in itself a fine achievement. But this good feeling can only be nurtured and sustained if some of the aims of his Agenda are fulfilled satisfactorily and soon enough.

Some of the items on the Prime Minister's Agenda are longer term and structural in nature and will therefore take more time to achieve. However, the government will have to take the necessary initiatives from now to solve these longer-term issues.

This has to some extent been already done with, for instance, the stepping up of the campaign against corruption and the appointment of the Royal Police Commission.

Several important persons have also been charged for corruption and this is a good sign.

Hopefully many more such initiatives will need to be followed up soon, in areas like the whole public service and in the private sector as well!

But there are nevertheless several other short-term items on the Prime Minister's Agenda which could be given greater priority to meet the public's expectations earlier.

These short-term measures will be those that are obvious in the public eye and must be addressed expeditiously. Some of them are as follows:

Firstly, all the counter services in the government departments, especially the long queues, have to be considerably improved, if the public is to believe that the government really means business!

This can easily be done, if the carrot and the stick are used simultaneously! The public wants to see tangible results, otherwise they will think it is all "No Action, Talk Only" (NATO)!

Government staff will have to become more polite and responsive to public enquiries. The public should not be indifferently pushed from pillar from pillar to post!

The government's delivery system has to be considerably and rapidly improved. This goal is not difficult to achieve if the civil servants, particularly those providing counter services, are more closely monitored for their daily productivity.

Those who exceed their productivity targets could be rewarded with overtime during extended hours of service to the public.

The long waiting time at queues could also be reduced if the government introduced 'Express Counter Services' to clients who would be prepared to pay higher charges for more expeditious and efficient services!

Furthermore, all those who do not want to perform as the Prime Minister's "team players" should be removed from the team!

Then the public will be truly convinced that the government really means business!

It is not difficult to identify the minority laggards and discipline them. This is only fair to the majority of civil servants who could be diligent and dedicated. The public wants to see more discipline and better service from government servants soon!

Secondly, another important public concern is that our very safety is now at greater risk. It is frightening that of late there have been an alarming rise in crime, murders, rapes, robberies and terrorist gangs!

The police force and other authorities and the NGOs will have to advise urgently, what is required, to act fast to prevent crime more effectively. We cannot allow our safety and security to deteriorate to the extent that, we are scared to walk our streets, and domestic and foreign business is scared away.

The Yang di-Pertuan Agong's concern over these rampant crimes has to be followed up with greater urgency.

Thirdly, The budget deficits have to be narrowed as a matter of priority. What is important is to build up surpluses on the Budget's current account, to help finance the development expenditures.

The Development Budget deficits are not necessarily bad in themselves, if the development projects are viable and have good socioeconomic returns. These would be projects that fulfil the basic needs for schools, hospitals, low-cost housing and clean water and better transport—especially for the poor. When this is done, the poor will believe that they are not doomed to poverty forever!

We should also not aim for high economic growth rates per se. Rapid growth can cause social evils that we are now facing!

In our further economic development, we should ask more carefully whether the projects are essential and whether they will benefit the citizens more than the businessmen.

Fourth, the ringgit exchange rate has to be kept under close review. The Second Finance Minister Tan Sri Nor Mohd Yakcop has rightly pointed out that Malaysia is keeping its options open on the currency peg. But to wait too long, for the regional currencies to fluctuate 20 per cent from the ringgit, or to wait for China to revalue, before we revise the exchange rates, may be just too late.

The business community must accept that they should not take the present fixed peg for granted, only because it provides them comfort, stability and albeit a false sense of security.

In the longer term, the fixed peg will make Malaysian businessmen complacent, and less competitive. They would have not taken the tough decisions necessary, to constantly re structure production and marketing, to face the stiff international competition posed by rapid globalisation. We have to be more proactive to avoid the economic shocks which we experienced in 1997!

Fifth, we need to increase the quality of our education and make it more accessible to all sectors of society—especially to the poorer sectors of our society. There also appears to be a wide gap between the improving school results and the actual quality and performance of our graduates! Hence we must find out how our school and university academic standards compare internationally? Why is there so much graduate unemployment? We should close this serious gap by raising our academic standards soon!

We could start now by giving higher budget allocations to our national primary and vernacular schools as well, to strengthen the education foundations, for a more qualitative rather than a quantitative education system. We should also insist that our teachers are better qualified and trained, so that they can be more productive.

Finally, the Prime Minister having set off on a flying start has now to keep flying high. The public will from now on monitor even more closely the quality of the election candidates and the government's success in attaining at least the short-term goals of the new Prime Minister's New Agenda. And, of course, we must also all work together with the new government to ensure that our country continues on the path of steady and sustained socioeconomic progress! Thus the new Prime Minister will have to offer a solid Election Manifesto.

The Barisan Nasional Manifesto 2004

The Chairman of the ruling party the Barisan Nasional and Prime Minister Dato' Seri Abdullah Ahmad Badawi, launched the National Front's General Election Manifesto 2004, on March 14 at the Putra World Trade Centre.

The Manifesto came one day after Nomination Day on March 13, 2004. The 15-page document had as its slogan the theme "Towards a Malaysia of Excellence, Glory and Distinction!"

The Manifesto also contained a kind of "Report Card" of all the major achievements of the government and carried "Pledges" to create "a better future for Malaysians and Malaysia"!

The Report Card and Pledges covered several vital topics and areas of major concern to the electorate, such as the economy, education, religion, law and order, public services and foreign policy.

Indeed, these are the basic issues which the voters are generally concerned with. The Barisan Manifesto has therefore addressed most of the pertinent issues of the day.

However, it is noteworthy that no mention was made about pushing the objectives of the NDP and particularly the "at least 30-per-cent Bumiputera ownership of the corporate sector"!

This is right and proper as there are many Malays and certainly more non-Bumiputeras who are tired of undue attention given to the NEP when the threat of rapid globalisation, is in fact more worrisome.

There is after all a major contradiction between providing excessive preferential treatment to *Bumiputeras* and at the same time wanting to provide them with the greater "competitive edge", to face up the real threat of globalisation!

On education again, it is important to enhance the opportunities for access to good education, rather than to spoon feed the less privileged to the extent that they will not learn how to acquire knowledge that will prepare them to compete effectively in the real world of competition.

In religion, law and order, public service and foreign policy, there must always be the "reality test" as to whether we want to seriously pursue the goals of Vision 2020 or to slide away from them?

We must move firmly on all these fronts in realistic and internationally competitive ways.

Otherwise, we might just be promoting a great deal of rhetoric which will sound good, but fall short of our national aspirations!

Indeed, the clarion call during the General Elections for "Excellence, Glory and Distinction" can become empty cries, unless we give real substance and meaning to them.

As it is, these goals do not seem very convincing. But we can and should make them solid, if we exercise determined political will and rally strongly around our leaders to realise these lofty aims.

We will have to wait and see how we succeed. That is why the Report Card on Performance is so important for our progress and success as a nation and as Malaysians!

Report Cards on Elected
Barisan Nasional Representatives

Dato' Seri Abdullah Ahmad Badawi's announcement on March 15, 2004 that Barisan Nasional's elected representatives will be assessed by a Key Performance Index (KPI) is music to the ears of especially the voters in the General Elections, on March 21!

It is a dream come true for all concerned Malaysians. Now the citizens will have Report Cards to assess the performance and quality of service provided by their MPs and state assemblymen and women so as to enhance the welfare and standards of living of our people.

The citizens will now be able to better monitor the delivery system and the benefits that were promised by the elected candidates, on a scientific and transparent basis, until the next elections!

The KPI will also encourage elected representatives and their officials to develop mission statements and specific targets, that have to be delivered effectively and on time.

The Prime Minister will also be able to see for himself how well his team is performing against acceptable benchmarks that cannot be shifted around like some goal posts.

The sound election promises and pledges for good governance are welcome. However, they will be of little use unless they are fairly implemented on the ground, with integrity and efficiency.

With the KPI, all civil servants and especially the elected political leaders will have to be fully accountable for their credibility—through their submission of quarterly reports to the government.

I believe that the Prime Minister's promise that "there is a place for everyone in this country" will be better realised, with the faithful implementation of the KPI.

The public could supplement the KPI by monitoring the performance of each and every elected representative at federal- and state-government levels, to ensure that all their election promises are indeed faithfully followed up and delivered on time.

We could also use the old and well-tested Red Book and Operations Room strategies that were adopted to fight communist terrorists during the Malayan Emergency (1948-1960).

Malaysia's second Prime Minister, Tun Abdul Razak, transferred this military strategy to fight poverty, particularly in the rural areas. The Red Book was kept in every district and subdistrict. It contained detailed information on the topography, population, and all the basic facilities and amenities available to the residents in that area.

Thus it was easy to keep track of the socioeconomic developments in the rural areas and to find out what the basic needs of the poor people were and how much progress was being achieved in meeting those needs.

It was a wonderful monitoring system that helped to break the back of poverty and to win the hearts and minds of the poor rural population as well as strong support for the government.

This same Operations Room and Red Book strategy could be used with of course more sophisticated computer technology, to monitor the performance of our elected representatives at federal and state levels and even at the grassroot level of local authorities, to serve the people more effectively.

I hope that the government will adopt sound monitoring systems to keep track of the performance of our elected representatives. The NGOs and civil society could also be encouraged to complement and supplement the government's efforts in this area of close supervision of performance and good governance.

Then we shall get closer to achieving our "Malaysian Dream"! But where are we heading?

Quo Vadis? After the Elections

The Barisan Nasional government won a landslide victory in the 11th General Elections on March 21, with a very impressive more-than-two-thirds majority!

Malaysian voters have given the Barisan government overwhelming support and a solid mandate to rule for the next five years!

But now the people will ask: Quo Vadis? What will be the Malaysian economic performance after the Elections?

I believe that the Malaysian economy will perform much better than before! But we must also work harder to place it on a better path for more sustainable progress in the future!

The economy will do better because the uncertainties as to whether Prime Minister Dato' Seri Abdullah Ahmad Badawi could lead the national coalition to win the vital two-thirds majority in Parliament have been removed.

From now on, the domestic and particularly foreign investors will feel more confident about our economic outlook and prospects. They will feel assured that the Prime Minister's business friendly attitude and cool non-confrontational style, will set a more positive tone for the new freely elected government.

Because of his strong mandate, in his own right, the Prime Minister will be able to take the tough decisions that are necessary to ensure greater stability, peace and prosperity.

It is thus essential that the government moves courageously and rapidly, to take the bull by the horns, so to speak, to fulfill all the ten promises that were made by the Prime Minister on January 31, 2004, just a hundred days after he took office.

People need to not only see but to feel the winds of change actually blow harder, before they recognise that they are going to benefit more from their new government.

But the good economic prospects cannot be taken for granted. Malaysia is only a small player in the vast global economy.

Bad Lessons from the U.S. and Australia

So we need to learn from the experience and lessons of more mature economies—and avoid their blunders! *For instance, the U.S. and its close ally Australia provide bad examples that we must not follow.*

Firstly, in the case of the U.S., we will have to very carefully watch its economic performance as it is our foremost trading partner that imports about a quarter of all our exports!

For all the euphoric economic forecasts of the U.S. authorities, including the U.S. Federal Reserve Chairman Alan Greenspan, the U.S. economy could be on a downturn from the later part of 2004!

But this questionable optimism is to be expected—with the U.S. Elections coming in November 2004!

The U.S. balance of payments deficit in 2003 is estimated at US$550 million or 5 per cent of its GDP, its budget deficit is at US$375 million or 3.5 per cent of the GDP. Its debt burden is large. The growth rate for 2005 year is thus realistically considered by some leading American analysts to be at only 2 per cent, down from the over 4 per cent estimated from 2004!

Hence we must avoid its economic pitfalls and be cautious about our trade and investment prospects with the U.S. economy that could slide!

On the other hand, the economic prospects with our other large trading partners like Japan and China appear to be far more promising.

Hence we need to give higher priority to developing our export markets with these countries and also with Singapore as it exports about a quarter of our products to other countries.

The U.S.-Asean Business Council in its latest Annual Report 2002, to the U.S. Administration and Congress, has recommended priority action for the U.S. to conclude the proposed Trade and Investment Framework Agreement (TIFA) with Malaysia.

While this move could be beneficial, we need to ensure that like most agreements with the rich industrial countries, we do not give too much and gain too little!

The U.S. businessmen do not seem to be concerned with getting the U.S. Administration to provide Asean and other developing countries fairer trade, like the removal of farm subsidies. The U.S. could lead in ensuring that it helps to reduce poverty in the developing countries through the adoption of more liberal farm policies, but U.S. businessmen do not seem to care!

The reduction of U.S. agricultural subsidies alone will thus contribute to the removal of one vital cause of international terrorism, because about half the world's population live in desperate poverty of less than US$2 per day!

Hence the recommendations made by the leading U.S. businessmen have to be viewed warily, so that we do not shoot ourselves in the foot.

In fact, Malaysia could be the largest importer of U.S. goods and services in Asean, since, unlike Singapore, we do not engage in much entrepôt trade as the little island republic.

Thus it is important that the U.S. Administration undertakes a more mutually respectful and beneficial relationship with Malaysia and other important developing countries. They should not want to trade with Third World countries, only to squeeze them for their own benefit, but go for fair and mutually beneficial trade

This way there would be more enthusiasm and confidence in building stronger, rather than weaker, trading relations with the U.S. And then there would be less suspicions of U.S. trade policies.

Secondly, the Australian economic weaknesses are also debilitating and we should struggle to overcome them in our country.

We have to ensure that we do not get into the "unacceptable and unsustainable" levels of poverty suffered by industrialised Australia, when we hopefully become an industrialised nation ourselves by 2020!

But we have to further strengthen our socioeconomic policies from now, in order to steer away from the depressing Australian situation.

According to a recent 470-page *Australian Parliamentary Report*, about 20 per cent out a total population of about 22 million Australians live in poverty on about A$400 per week. This low income compares unfavourably to the minimum Australian wage of A$431 per week!

Australian Prime Minister John Howard dismissed the study saying that recent changes in welfare payments had greatly benefited the poor. But Senator Steve Hutchins chairman of the Senate Committee and a member of the Opposition Labour Party said "Our nation is rich and great, but that children are still going to school hungry is unacceptable"!

There is no point in achieving industrial nation status, if our economic management is inefficient or the income gap between the rich and the poor in our country, gets so skewed!

Hence the real challenge facing the new Malaysian government and Prime Minister is to develop policies and strategies that would enable Malaysia to achieve the goals of Vision 2020, without the structural socioeconomic weaknesses of the U.S. and Australia.

To follow their paths would lead to disunity and instability in the longer term. *Malaysia must therefore steadfastly continue to strive and indeed "struggle" for national unity, peace and prosperity that will be acceptable and sustainable for all Malaysians for now and the future.*

The Winds of Change towards the Malaysian Dream?

Prime Minister Dato' Seri Abdullah Ahmad Badawi has now been leading the country for about six months. It is also over a month since he won a massive victory at the General Elections and formed his Cabinet soon after.

Hence have there been the expected winds of change and how strong have they been? Indeed, the winds of change have been blowing with all the Election pledges and promises—and the economy is sailing in the right direction.

But although several sound policy initiatives have been introduced, they have already raised serious concerns about the quality of their implementation.

Examples of Poor Implementation

Let's take a few striking examples as follows:

1. **The Election Commission has made several technical blunders that have cast a cloud over the otherwise successful General Elections.** Why can't the public be informed of at least the preliminary findings of the investigations into the election anomalies.

237

2. **The National Service programme appears to have been hastily implemented without sufficient planning.** Hence we have so many unnecessary weaknesses and heartaches that could have been easily avoided. The authorities would need to come out quickly with solutions to obvious shortcomings, rather than wait for a full study to be completed.

3. **The Royal Police Commission, despite its noble intentions, does not seem to have earned adequate public feedback on how to improve the police force.** We need to find out why? Is there a lack of public confidence in the Police Commission because of any fear to complain against the police? We have to find out from NGOs and the public why the support is slow in order to make the Police Commission more effective.

4. **The establishment of the National Integrity Plan is most welcome.** However, we must ensure that the newly formed Integrity Institute Malaysia will be properly staffed and be able to really deliver honest service to the public. There is little advantage in setting up institutions and then not doing much that is concrete and beneficial to the public because of ineffective staff.

5. **Then again there is the embarrassing poor passing rate of only 0.9 per cent for 130,000 teachers taking the Assessment of Efficiency Level test.** This low passing rate could have been much higher if the teachers had been consulted in the design of the examination questions. The teachers should have been tested on their teaching competencies, rather than their knowledge of administrative and financial regulations!

Hopefully Dato' Hishammuddin Hussein Onn will cancel the test and order a revamp of the structure of the test to make them more meaningful in assessing the real quality of our teachers!

If all new initiatives like the ones mentioned above are not properly implemented, there could be a growing credibility gap, which could become economically counterproductive.

It must also be appreciated that our ability to strengthen the winds of change will depend on our capacity to meet the growing international economic challenges of a possible slowdown among the world's richest countries.

However, our pace of progress and ability to keep on course, will depend on our ability to face the growing international competition and our successful implementation of present and any new policies.

Thus we also need to be forewarned that there may be more constraints on our ability to introduce changes.

The government will have to develop a stronger momentum in introducing the winds of change.

CHAPTER 12

STRONGER WINDS OF CHANGE NEEDED

WHAT we need to do now is to revise our economic and financial planning, to find out what new strategies we can devise to counter the adverse effects of any international economic slowdown and to encourage our own domestic economic growth?

The structure of the present Economic Planning Unit (EPU) has not changed much since it was formed soon after Independence. We need to redesign the EPU and ensure that it is more competent to deal with the latest challenges to economic planning, to face globalisation.

IMP3's New Directions

Our international trade structure should also change. Dato' Seri Rafidah's announcement that MITI is working on Malaysia's Third Industrial Master Plan, 2005-2020 (IMP3) is most welcome.

The IMP3 will seek to accelerate the development between "manufacturing-related services and enhance the development clusters".

For this reason the role of the successful Malaysian Industrial Development Authority (MIDA) has been expanded to include the Manufacturing Services Division, to further develop the service sector.

However, I hope that this is only a transitionary policy measure. It would be more effective if the service sector (which contributes about 56 per cent of the GDP), is given a separate and more specialised agency, such as a proposed Malaysian Industrial Services Authority (MISDA) to accelerate the progress of the service sector?

The manufacturing sector grew by about 10 per cent in 2003 while the service sector expanded by about half the pace at only 5 per cent. This is because the service sector exports amounted to only RM1.5 billion while its imports accounted for a huge RM5.9 billion.

These startling figures indicate the crying need to give the service sector a much greater priority and push. MIDA will need to be considerably restructured and other vital agencies like the EPU and the Treasury will have to be reorganised to pursue a more integrated strategy, to really strengthen the service sector as a dynamic growth area.

The related manufacturing services of banking, telecommunications, education, construction and professional services like accountancy, legal, engineering and medical, have to be rapidly expanded and enhanced in quality. And we must have new and more dynamic planning to achieve these higher goals.

But is our education system sufficiently geared to meet these international challenges of greater global competition?

Most Malaysians would have problems with the present education system. They will readily agree that it has to be considerably improved, to strengthen national unity, and to enable our graduates to be more marketable.

Education can also help to reduce poverty and especially "hardcore"poverty, if students are taught the basic R's of 'Reading, Writing and Arithmetic' more effectively.

Poverty Eradication: New Approaches

The government's newly announced policy to eradicate hardcore poverty within 3-5 years in the rural and urban areas respectively will be most welcome by all Malaysians

But why should there be different target dates for eradicating hardcore poverty in the rural and urban areas?

I hope the government will use the same target date to eradicate hardcore poverty in both the rural and urban areas, without any differential treatment.

It is also reassuring that the government wants to review the present poverty line for hardcore poverty which is less than RM529 per household of 5 persons, per month in Peninsular Malaysia! This works out to about RM106 per month or about RM3.50 per day per person, for food, shelter, clothing, transport, etc. But how does anyone manage to live on even basic needs, on this small sum please?

However, even with a more sustainable poverty line of say RM750 per household per month, I believe that poverty and particularly hardcore poverty can be eradicated, if there is stronger political will and if the full cooperation of civil society is obtained.

Hardcore poverty is defined as half the poverty line income of RM529 per household per month in Peninsular Malaysia. This is absolute poverty with which must be extremely difficult to even subsist! Surely we have to go all out to eradicate hardcore poverty, to maintain social stability, in our national interests!

Both Dato' Seri Ong Ka Ting and Dato' Abdul Aziz Shamsuddin could draw on the policies and rich experience that is already available in the government and among the NGOs. They could therefore accelerate the implementation of many of these policies immediately.

The rate of achievement in eradicating not only hardcore poverty but poverty in general, could provide good examples of how successfully the General Election pledges are being implemented.

Poverty is one major root cause of social unrest, national instability and even international terrorism.

We must all therefore rededicate ourselves to this noble task of raising the standards of living of our poor and of especially our hardcore poor, as a top national priority.

Then we will achieve *Bangsa Malaysia*, and strengthen our socioeconomic resilience and national unity.

Dato' Abdul Aziz Shamsuddin is to be commended for taking the initiative to announce his target to modernise rural villages in just three years!

He has asked his officials to prepare plans and activities in three months that would reactivate and make the his Ministry more dynamic!

Indeed this would hopefully be the kind of ministerial thrust that all ministries and departments could adopt to great advantage. This would be the basis for monitoring the serious pledges and high performance that was promised during the recent election campaign.

I hope that the eradication of hardcore poverty will be equally applied to the large numbers of the urban poor of all races, who have received far less priority in the past.

That is one reason why we had riots at Kampong Medan in the heart of Kuala Lumpur. We must thus avoid this dichotomy in our treatment of the very poor marginalised groups at all costs—regardless of race!

Civil society would need to be more actively supported
to complement the government's efforts to eradicate poverty
as the government cannot do it all by itself. The government
often cannot provide the most effective cost benefits in its
capital and human investment to improve the welfare of our
vulnerable groups. Hence the NGOs could help a great deal
to combat poverty.

*Malaysia is well ahead in our commitment to meet the U.N.
Millennium Development Goals. However, as an advanced
developing country, we could aim for "Millennium Development
Goals PLUS", since we have the experience and the means to
break the back of poverty sooner than other developing and even
some industrial countries.*

Hopefully Dato' Abdul Aziz and Dato' Seri Ong Ka Ting
will bring back the Operations Room techniques and
harness the great energies of civil society, to improve the
regrettably poor delivery system. Poverty breeds some of the
root causes of social unrest and has to be countered at the
grassroots.

We must all therefore dedicate ourselves to this noble
task of raising the standards of living and the quality of life of
our third class and all under-classed Malaysians, as a matter
of urgency. Only then can we achieve *Bangsa Malaysia*,
strengthen national resilience and our continuing prosperity.
But there is much doubt that the education system is not
sufficiently geared to meet the greater international
competition for which we need employable graduates.

Unemployed Graduates
and Poorly Paid Doctors

There is thus growing concern that the ranks of the
unemployed graduates are increasing. They are mainly
Malay graduates from the public universities who have

unfortunately been denied adequate studies in the English language.

The attachment of 27,000 unemployed graduates to industries will help them eke out a living. However, this programme will not solve the problem of enabling them to earn what their peers from the private colleges and universities will earn.

Hence some radical and urgent measures have to be adopted by the new Minister of Higher Education Dato' Dr Shafie Mohd Salleh to prevent the queues of unemployed graduates from getting any longer.

A similar crying need for change is evident in the case of government doctors. They earn a paltry salary about RM3,500 per month after about 10 years of selfless service and are paid a miserly RM1.04 per hour for overtime on weekends! No wonder they quickly resign and cause a major shortage of doctors in the public sector.

Would the new Minister of Health Dato' Dr Chua Soi Lek help to rectify these serious anomalies as an emergency? Housing—like health—needs greater priority.

Low-Cost Housing

The Selangor government's proposal to sell low-cost houses to persons with higher incomes is a partial solution to the housing problems.

But the Cabinet rejected the decision as low-cost housing is provided for low-income groups and not for those with higher incomes!

The problem of severe shortage of low-cost housing has been a sore problem for thousands of low-income and houseless households for a long time.

Developers are required to reserve at least 30 per cent of their housing units for the low-income groups. However, these units are often not taken up by the target low-income groups since they usually do not have the means to book and buy these low-cost houses.

Hence the developers generally face financial problems unless they have high reserves. They have to service their bank loans without sufficient income flows from the sale of these low-cost houses, and hence some of them get into trouble with the banks.

Thus the developers can therefore get into financial difficulties and sometimes even have to abandon their housing projects, if their houses are not sold.

So it makes sense for these low-cost houses to be sold to those higher income groups who could be required to rent these houses at reasonable prices to the low-income groups with some provision to buy back the houses at an appropriate time.

But the Cabinet by its decision to deny the sale of these houses to the higher income groups, does not solve the problem of the developers. Neither does it help the low-income groups to rent and perhaps also buy these low-cost housing units in the future.

The real solution would be for the government to make much larger allocations for subsidised low-cost housing, or to buy the completed housing units from the developers or preferably build more low-cost housing units itself from the poor!

Another solution is to provide incentives for housing developers to undertake Industrial Building Systems that will lower the cost of building, improve standards of building, hasten the building processes and thus save costs of borrowing.

The Industrial Building Systems could contribute substantially towards solving the serious housing problems for the low-income groups in our society! But there is resistance from the industry! This is because vested interests are affected.

According to the Real Estate and Housing Developers' Association (REHDA Malaysia), a million families earning between RM1,500 and RM3,000 per month are ineligible for low-cost housing. Only families who earn less than RM1,500 per month are eligible.

Minister of Housing and Local Government Dato' Seri Ong Ka Ting has promised to review these guidelines for low-cost housing. It is hoped that the government will be able to improve the low-cost housing situation soon as it has been festering for too long! More beneficial policy changes are required.

The Winds of Change
Must Gather Momentum

The expected winds of change have to gather more momentum if the people's high expectations are to be reasonably fulfilled.

However, this means that socioeconomic planning and policies need to be fully reviewed and revised—and even restructured, if need be—in order that we can progress more purposefully. But we have to be careful too that we do not bite more that we can chew, lest we get heartburn and indigestion. New policies can help but more damage can be caused by a lack of implementation capacity or the political will to strengthen the winds of change!

Although the winds of change are being felt, they should blow at a faster and more steady pace. Prime Minister Dato' Seri Abdullah Ahmad Badawi enjoys the high confidence and much support from the vast majority of Malaysians, for his policies to take Malaysia forward.

Malaysians will work steadfastly "with Abdullah" as he has requested. However, because their faith and trust in him is so significant, they also have "great expectations" for him.

We can only hope and pray that our Prime Minister will be blessed with the strength and stamina to lead Malaysia and all Malaysians to greater glory in the future.

Indeed, we could be a model multiracial, prosperous, harmonious and modern Muslim nation—if we all contribute wholeheartedly and sincerely to making Malaysia really great!

God willing, we pray that the winds of change will blow more strongly so that we will succeed in achieving the reality of the Malaysian Dream!

Addressing Outstanding Issues

The Malaysian Treasury has stated that it will keep to its earlier projection or estimate of a 6-6.5 per cent growth rate for the economy in 2004. This is encouraging considering the fact that the U.S. could increase interest rates and that oil prices have gone up. Furthermore, there is the deteriorating security situation in the Middle East, especially in Iraq, and increasing international terrorism.

Hopefully, the Malaysian economy is sufficiently decoupled for us to be unaffected by any slowdown in the U.S. economy and even globally.

Given these uncertainties, I would think that the private sector should be more cautious and better prepared for any contingencies that may arise. It is thus important that we address our outstanding national issues more urgently in order to strengthen our economic resilience.

1. **Non-Payment of Taxes.** It is surprising that the chief of the Inland Revenue Board (IRB), Tan Sri

Zainol Abidin Rashid, has stated that the ten top income tax dodgers have not paid taxes amounting to some RM1.0 billion! This figure would be much higher if the total unpaid taxes is taken into account. Perhaps Zainol should be more transparent and announce that fascinating figure, too?

This welcome revelation, however, raises the important question in the minds of all taxpayers and especially the smaller taxpayers: why and how these rich taxpayers have been allowed to get away with not paying such large debts to the IRB for so long? Why did the IRB not catch them or make them pay up earlier before the tax debts piled up so high?

The smaller taxpayers who dutifully settle their taxes will feel more comforted if they are given more explanation as to why these big fish have got away so far and what will be done to collect the tax debts for economic growth and to reduce the high Budget deficits!

2. **Declaration of Assets.** It is alarming that a large number of civil servants have shown their defiance in not declaring their assets on time, despite having been given four months to do so! Many have even disregarded the additional grace period of two weeks up to April 15, 2004.

    But there are also many civil servants who have shied away from declaring their assets, although disciplinary action will be taken against those who have not declared their assets before the extended date of May 15, 2004!

    This is a shame as it reflects on the state of indiscipline in the civil service and the disregard for even the orders from the head of the Public Service

Department, Tan Sri Jamaluddin Ahmad Damanhuri!

It is therefore hoped that the government will take a serious view of this insubordination and will take stern and early disciplinary action against these recalcitrants in the civil service. Perhaps there would be a salutary effect on the civil service if the names of all those who have not met the deadline could be published and embarrassed into declaring their assets.

3. **Transfer of Civil Servants.** The announcement by the Director-General of the Public Service Department (PSD) that there will be a massive transfer of civil servants from sensitive sectors is welcome. Those who have served 3-5 years in sensitive departments like the Immigration, Customs and Land Offices, where the temptation for corruption can be greater than other places, will be transferred.

    This is a clever way of reducing corruption and inefficiency without having to find proof of distortion in implementing government policies and of the abuse of financial trust. Furthermore, disciplinary action against civil servants are notoriously tedious and long drawn. Hence this is a more practical and fastest way of dealing with errant civil servants.

    However, I hope the PSD will not be passing the buck and will nevertheless take firm discipline against those derelict civil servants who deserve to be penalised or even dismissed.

4. **Budget Consultations.** Prime Minister Dato' Seri Abdullah Ahmad Badawi's statement at the Budget

Consultations 2005 that "We are still far from hoping to transform our country, judging by the number of applications submitted by industries for unskilled foreign workers" was most telling.

The government could encourage more capital-intensive production if it is not so willing to allow the private sector to easily import low-wage foreign workers.

If the government is more restrictive in the import of foreign labour, the construction and plantation industries would be compelled to undertake more research in using more labour-saving methods.

This move would also help to raise wages for our local labour and thus reduce poverty and raise the standards of living of our own workers.

MTUC President Zainal Rampak expressed concern on Workers' Day that despite satisfactory economic growth rates, some companies were laying off workers through Voluntary Separation Schemes (VSS).

On the same occasion, the Executive Director of the Malaysian Employers Federation (MEF), Shamsuddin Bardan, stated that "we have to shift from the current wage structure of automatic annual increment and fixed contractual bonus, which has no regard to the performance of employees and profitability of companies".

Both concerns are valid. *Perhaps the best solution is to reduce foreign workers, allow local wages to rise to market levels and then agree to relate wage increases to productivity increases, in the spirit of smart partnership.*

But the employers should also provide more training to raise the productivity of the workers who are ready to acquire higher skills.

Unless we take urgent action to raise productivity and give workers a fairer deal, we will not be able to maintain the high level of industrial peace nor be able to compete effectively under globalisation.

5. **Industrial Coordination Act 1975 (ICA).** The Federation of Malaysian Manufacturers (FMM) has once again asked the government to raise the Industrial Coordination Act (ICA) threshold, from RM2.5 million to RM10 million.

When the government wants to strengthen the performance of small and medium industries (SMIs), this outdated threshold appears as a serious contradiction and hopefully the low ceiling will soon be raised.

However, the FMM has also been urging the government for a long time to reduce the corporate tax from 28 per cent to 22 per cent in 5 years. But this request does not suggest how the loss in government revenue could be compensated. At the same time the FMM keeps asking the government to improve and expand the infrastructure facilities to reduce the cost of doing business.

The government cannot accept both proposals at the same time and also reduce the accumulating Budget deficits and the FMM knows it! Budget proposals from the private sector must be reasonable and credible!

6. **Too Few Patents.** The Minister of Science and Technology and Innovation, Dato' Dr Jamaluddin

Mohd Jargis stated at the ASLI Conference on "Rethinking Malaysia" that only 7 out of 1,000 inventions in Malaysia are patented as compared to 220 in Japan, South Korea and Taiwan.

This is indeed a very low achievement in Malaysia. However, it begs the question as to what should be done to increase our patents and to raise our standards of science, technology and innovation?

Priority has to be given to the teaching of science and technology in our schools. More time and resources to teach science will help. But we should also raise the incentives to study the more demanding subjects like Mathematics, Physics, Chemistry and Biology.

Perhaps students should be encouraged to study both Science and Liberal Arts in our colleges and universities. Then there would be a wider and stronger base of science graduates who would be able to do more research and obtain more patents.

Furthermore, we should rethink and ensure that the government at least should raise the salaries of such scientific personnel as engineers, scientists and doctors.

We need to also ask why so many young professionals plan to migrate so that we could reduce the brain drain, instead of trying to brain gain by enticing Malaysians from abroad by allowing them to import two cars.

7. **Competitiveness Ranking.** On the positive side, it is heartening that Malaysia is the 16th most competitive nation. Our efficiency rose five rungs compared to 2003, according to the Swiss-based

IMD, which is an independent non-profit foundation.

However, we performed less creditably in promoting private investment, developing new sources of growth, enhancing efficiency in the delivery system and enhancing the quality of human resources.

Many of these challenges to improve our competitiveness stems from the unsatisfactory implementation of the NDP. However, this is a sensitive and difficult area that has nevertheless to be carefully managed if we are to become more competitive and at the same time maintain balanced growth and income distribution.

The way to achieve this vital balance is to help the poor of all races under the new development policies, but to insist on equal treatment and meritocracy, once students have graduated!

8. **S&P Rating Rises.** The latest Standard and Poor's (S&P) rating of Malaysia gives an A-/A-2 for our foreign currency and a rating of A+/A-1 for our local currency, with a stable outlook!

This is an impressive rating, especially for a developing country. But we can improve on it if we take greater concerted efforts to overcome some weaknesses that S&P has outlined as follows:

i. the continuing Budget deficit for 6 years; and

ii. the rising net debt of the government which has grown to 48 per cent of the GDP. This is well above the "A"-rated median of 26 per cent!

It is therefore important that the government gives much higher priority to reducing the Budget deficits at a faster pace.

The private sector then should be given sufficient incentives and encouragement to expand at a faster rate to compensate for the cutbacks in government expenditures. The private sector should be encouraged to play a larger role in economic development if we push forward with more privatisation—provided, of course, that the poor sections of society do not lose out in the process!

9. **Khazanah Nasional.** The restructuring of Khazanah Nasional and the transfer of 30 government-linked companies (GLCs) from the Ministry of Finance Incorporated to Khazanah Nasional will raise the efficiency of these GLCs and strengthen the Malaysian economy.

    Historically many of these GLCs were managed by civil servants who had no real track record in business. The goal was to deliver good essential services to the public while performing a social role in providing employment for *Bumiputeras*.

    Consequently, the GLCs were not very efficient. As a CIMB Securities report pointed out, the efficiency improvements of only five GLCs—Telecoms, Tenaga Nasional, Sime Darby, Golden Hope and Kumpulan Guthries—would result in a net profit of about 20 per cent. This would raise the market capitalisation by RM20 billion, or 5 per cent of the KLCI market capitalisation!

It is estimated that 40 GLCs have a combined market value of about RM232 billion or 34 per cent of the total market capitalisation of Bursa Malaysia.

This does not include the stakes in GLCs held by Permodalan Nasional, the Employment Fund and Petroliam Nasional.

If all these huge holdings by *Bumiputera* institutions are taken into account, the *Bumiputera* equity share of the corporate sector of the economy will exceed the current estimate of 19 per cent and could probably be in excess of the national target of 30 per cent!

The Second Finance Minister Tan Sri Nor Mohd Yakcop is spot-on in wanting to heighten the professionalism in the management of the GLCs to increase productivity and profits.

These innovative moves will enhance economic growth and the earnings of stakeholders, including that of the ordinary shareholders.

These measures are only self-evident and should have been adopted long ago, if not for the past tendency to tolerate inefficiency as long as the social needs of the NEP were met!

With the use of Key Performance Indices (KPIs), it is hoped that the scrutiny would be rigorous, otherwise taxpayers will be paying more for the same output.

This would be even more wasteful and more damaging to the economy. The taxpayers will therefore be watching more carefully for the more transparent results of these new GLCs!

But all this begs the question as to what is to become of the senior civil servants?

They will continue to advise government Ministers on policy, some of which will relate to the GLCs.

But how much respect can the top-notch, highly paid managers of the GLCs give to the civil servants (whom they have to report to) if the civil service is increasingly staffed by less qualified and relatively less remunerated civil servants?

Should the government not also revamp the civil service and pay them market rates and reward them on the basis of performance too?

While it is justified in appointing some of Malaysia's best business brains to serve on the Board of Khazanah, it will be a pity if all the best brains come only from one ethnic group. The custodians of public finds in Khazanah would need to be representative of all Malaysian ethnic groups to enjoy credibility and to promote national unity. Civil society can influence government to have more ethnic balance.

Civil Society Contribution

Civil society can contribute substantially to better public policy formulation and improved implementation if the NGOs carefully monitor the performance of the privatisation projects to ensure fair pricing and good quality of the deliverables.

The Cabinet decision to form a Select Committee to examine the draft amendment to the Criminal Procedure Code is therefore most welcome, particularly when the committee is to be made up of government and even Opposition members and NGO representatives!

This healthy development of a smart partnership among government, the Opposition and Civil Society must be strongly encouraged.

It is gratifying that the Prime Minister informed the 30 NGO groups which attended his Budget Consultations that "NGOs function as a check and balance to inform government about its shortcomings in the administrative system".

This gives a stronger backing for the important role of NGOs in developing our society. It also signals to the public service that they have to give more attention to the views of NGOs whose ideas have often been regarded with a degree of indifference. NGOs need support from all, including royalty!

The Selangor Sultan's Directive

The Sultan of Selangor Sharafuddin Idris Shah rightly issued a directive in his Royal Address to the 11th State Assembly that the state government should come out with workable action plans, to resolve the urban problems of squatters, the dirty environment and the poor maintenance of public buildings.

But will the State Assembly treat the directive seriously and urgently?

The caring sultan also expressed his dissatisfaction with cleanliness of public amenities, such as markets, playing fields, food stalls and restaurants and the rising crime index.

However, will the state government and local authorities take up the challenge of solving these problems?

I believe that the Mentri Besar will respond favourably by urging the relevant state authorities to do something about the sultan's complaints in the short term.

However, these knee-jerk responses will not be sustained as the state authorities and their staff will not have the stamina and the will to carry out the considerable efforts needed to continue with good governance.

What is needed therefore is for the state government to monitor the implementation of follow-up action and the proper implementation of state government policies if the sultan's sound advice is to be given the respect it deserves!

Where it is found that the political leaders and civil servants are wanting, the sultan could pull up the Mentri Besar and the State Secretary and take the necessary disciplinary action— and to keep the public informed of its follow-up action on the sultan's royal directive!

Then we can let the citizens decide whether the political and civil service leaders should continue to stay in office at the next Elections!

I am sure the citizens will hope that other sultans will take the lead in demanding good governance in their respective state governments and in national policies, including university intakes!

University Intake

The government has made commendable progress in improving the quantum and racially balanced student intake into the public universities in 2004. But the present meritocracy system is only partial and could be further enhanced. For instance, we need not have both the STPM (*Sijil Tinggi Persekolahan Malaysia* or Higher School Certificate) and the Matriculation streams, but only one stream in the future. Then we will have real meritocracy!

It is very revealing that of the top scorers, only one *Bumiputera* student out of 527 top STPM scorers gained admission to the universities!

Furthermore, the courses offered at KUIM, UIAM, UiTM and the Islamic studies at UM and UKM should also be included, to give a full picture of the *Bumiputera* share of the total public university intake, which then could be well above the present 63.8 per cent.

The university intake of Chinese students declined from 11,921 in 2003 to 11,778 or by 1.3 per cent to 30.3 per cent, while the Indian intake rose slightly from 1931 students in 2003 to 2,277 or by 0.7 per cent to 5.9 per cent in 2004.

These figures appear to reflect the racial composition of the total population and will be more acceptable to the general public.

Although the Chinese intake is less than in 2003, it has to be borne in mind that this does not include the large intake by the MCA-owned University Tunku Abdul Rahman (UTAR). The Indians intake also does not include the Indian student intake into the MIC-owned Asean Institute of Medicinal Sciences (AIMS).

Unfortunately, in trying to achieve the goals of meritocracy and racial balance, many top students had to be deprived of places in the critical courses which they chose, like Medicine, Engineering and Law.

Malaysia cannot afford to under-utilise its human resources and erode its economic resilience and national unity and accentuate the brain drain through poor quality university intakes.

Proposals to Increase University Intake

Hence we need to refine the present university selection system with, *inter alia*, the following proposals:

1.   **Increase the capacity of public universities, especially in the critical courses, like medicine**

engineering and law. The reason given that medical places are constrained because of a shortage of lecturers is difficult to accept. There are many top medical specialists in government and especially in the private sector who would gladly offer their teaching services on a part-time basis.

2. **Raise the academic standards of the STPM and especially the less demanding Matriculation examinations, so that we do not get a huge number of 1,774 students who obtained a maximum Cumulative Grade Point Average (CGPA) of 4.0!**

   It must appear very unfair to the top students not to get a university place of their choice—like medicine. Thus the 128 students with the maximum CGPA of 4.0, who wanted to study medicine as a first choice, did not get places in medicine!

   I hope they and other high scorers will gain admission or given scholarships to study elsewhere!

3. **Introduce full meritocracy within each racial group rather than continue with the present partial meritocracy among all the racial groups.** This new approach will help to achieve a more balanced representation, according to the racial composition of the Malaysian population.

4. **Encourage the expansion of private universities and colleges to absorb those eligible students (who could not get admission to public universities) into private education institutions with government-subsidised fees.**

   As it is, only 45 per cent of the eligible students of 38, 892 students, with a minimum requirement of 2.0 CGPA, got places in the 14 public universities.

   Hence there is a vast potential to increase student enrolment in private institutions of higher learning. Technical education could also be easily expanded to meet this strong demand for tertiary education!

The provision of more scholarships or loans to students to enter private tertiary institutions of higher learning, will also increase the number of *Bumiputera* students in private colleges and universities and thus help promote increase national unity.

5. **Provide more priority teaching and educational facilities to the poor and bright students, and give them preference in admission—as intended by the NDP!**

In the rural and urban poor and depressed areas, more attention should be given to upgrade their opportunities for educational advancement and university entry.

Thus the public university student intake, can be considerably improved, if the present selection system is reviewed, to take into account the above and other suitable criteria. There should be more public debate and adequate feedback to the government, which could then revise the present policies and practices, to increase public support for the university selection system. Malaysia will then become even more harmonious and more internationally competitive.

Weak Graduates

The new Minister of Higher Education Dato' Dr Shafie Mohd Salleh stated very clearly at the his first meeting with the 17 vice-chancellors in May 2004, that "the mismatch between jobs offered and skills demanded in the market deters graduates from getting the right job"!

He is right as is evidenced by the 21,000 unemployed graduates as in May 2004. These unemployed graduates are mainly from the Social Sciences and Islamic studies. These graduates are simply far less marketable.

Part of the problem is that undergraduates appear to want to choose the 'soft' optional subjects, even in preference to the 'harder' core subjects.

Thus in a recent survey commissioned by the government only 57 per cent of the graduates got employed, while 29 per cent were jobless and about 14 per cent went on to do further studies—many of them because they could not get employed!

Hence the Graduate Training Scheme has been introduced to expose graduates to work on attachment with industry. This scheme appears to be quite successful as about 74 per cent of the 12,449 graduates in the scheme found employment in the industries to which they were attached. The rest still could not get jobs soon after the training and had to wait longer for employment.

Finally the motto of the High School Bukit Mertajam which is rightly proud of its most illustrious graduate Dato' Seri Abdullah Badawi, states "Accomplish or do not begin"! I have no doubt that the Prime Minister will accomplish the transformation of the Malaysian economy to enhance our peace, stability and national unity to benefit all Malaysians—regardless of race.

Then "the winds of change" that have been so well introduced by the new prime minister Dato' Seri Abdullah Ahmad Badawi, would have achieved our aim to build a peaceful, progressive and prosperous Malaysia for all Malaysians. However, we have to overcome the many challenges facing the country.

CHAPTER 13

CONCLUSION:
THE CHALLENGES FACING
MALAYSIA IN THE FUTURE

MALAYSIA will face many challenges on several fronts in the next few years. However, its potential and prospects in overcoming them are encouraging, provided that Malaysian leaders continue to be strong and determined to make Malaysia succeed. This favourable outcome is highly probable.

But what are some of these major challenges?

The first major challenge is to continue to strengthen Malaysia's national unity. This can be achieved by maintaining its political stability and attaining higher economic growth and better income distribution.

The second big challenge would be, how to face up to the threats and opportunities of growing globalisation? But there are many other lesser challenges as well.

Indeed, the next few years will see the evolution of the Malaysian political economy, after the transfer of leadership from the charismatic Prime Minister Dato' Seri Dr Mahathir

Mohamad to his able successor Dato' Seri Abdullah Ahmad Badawi.

With Dr Mahathir's retirement at the end of October 2003, after about 22 years as a dynamic, innovative and sometimes controversial Prime Minister of Malaysia, Abdullah assumes the country's top political seat in the smoothest and also the longest leadership transition the country has ever seen.

Dr Mahathir's leadership has been outstanding in many ways. History will judge him as the *Father of Modern Malaysia*, although no leader can have a perfect record.

The immediate leadership challenge therefore is for Abdullah to carefully help to steer the ship of state well after he takes over the reins of government.

He has to steadily consolidate his new leadership, to build upon the many achievements of Dr Mahathir and to further improve the management of Malaysia's political economy. There is much confidence that Abdullah Badawi will lead Malaysia impressively.

Malaysia is one of the most complex political economies to lead and to manage, given its significant multiracial, multireligious and multilingual mix. Nevertheless, Malaysia's natural resources and well established institutions, enable the country to have promising prospects to continue to progress and to proper.

Malaysia is now one of the most advanced developing countries and arguably a rapidly emerging developed country. This is evident with its relatively high standards of living, with income per capita of about US$3,500, a good quality of life, strong government, and genuine peace and stability.

It has some world-class projects like the tallest building in the world—the Petronas Twin Towers, the Kuala Lumpur International Airport, and the country's many First World infrastructure facilities. Nevertheless, the challenge is to provide more highly qualified personal human resources, to sustain the high growth rates of the Malaysian economy.

Malaysia aspires to be a fully developed industrial country according to all the acknowledged norms, through its Vision 2020. It is most likely to achieve that aspiration by 2020. In fact, Kuala Lumpur and the plush new city called Putrajaya which together constitute the Federal Capital, together with the state of Selangor in the affluent Klang Valley, could qualify even now for industrial-nation status!

National Unity Weakening

The challenge of building greater national unity will have to be given higher priority because in recent times unfortunately, polarisation appears to have increased rather than decreased in most sections of Malaysia's multifaceted society.

The main cause of polarisation is found in the education system which provides for separate primary schooling in the country's main language streams—Malay, Chinese and Tamil—as provided for in the country's Constitution.

The government primary schools are predominantly Malay, in student enrolment and teaching staff. The same is true for Chinese primary and secondary schools which are attended primarily by Chinese students and teachers. This is also the case with the Tamil primary schools. Hence there is minimal mixing of the Malaysian students of different races from childhood until the secondary school level. By this time, some polarisation has already set in and carried on at tertiary levels of education as well.

The challenge therefore is to find a reliable formula that will be accepted by all races whereby, primary school students can be taught in Malay, English and in their mother tongues like Chinese and Tamil. The existing Islamic religious schools where the teaching of Islam is emphasised makes it more complicated.

The government had proposed the setting up of Vision Schools to meet this challenge of polarisation, but full community support has been lacking.

Thus the students at government universities and institutions of higher learning have become less nationalistic in their identity. The private colleges and universities, on the other hand, have mostly Malaysian Chinese and foreign students. All these trends undermine national unity.

This racial compartmentalisation is also found in the public service, including the teaching service and the armed forces, which are largely staffed by Malays, especially at the higher levels of policy formulation and management.

Similarly the business sector is mainly owned, managed and staffed by the Chinese who own the largest share of the corporate equity in the economy. Hence the government has undertaken socioeconomic restructuring to bring about better ethnic balance in the ownership in the corporate sector.

The Malay-dominated government owns and manages the "commanding sites" of the economy, such as petroleum, the utilities, ports, the national airline, railways and all major infrastructures. However, the mainstream commercial and business activities are managed by the minority Malaysian Chinese and not by the majority Malay or *Bumiputeras*, including the ethnic groups from the East Malaysian states of Sabah and Sarawak.

The Indians are employed mostly at the lower levels in both the government and the private sectors, except in the professional groups in the private sector where they still have a higher proportion of representation, compared to their racial composition. But this favourable proportion is declining too.

The position of the Malays has improved significantly in the business and professional sectors under Prime Minister Mahathir's leadership and the active promotion of the affirmative-action policies. But the challenge to narrow the racial gaps to achieve national unity and to eliminate the identification of race with occupation will unfortunately remain for some time to come.

Affirmative Action: Review?

Affirmative-action policies have been implemented since the serious racial riots and social unrest that shook the nation in May 1969.

These affirmative-action policies (also known as the New Economic Policy and subsequent variations of it) have contributed greatly to the peace, stability and remarkable economic growth and prosperity that Malaysia has enjoyed since Independence in 1957.

Malaysia's overall progress has been all the more impressive when compared to most, if not all, other developing countries.

But the challenge now after about 35 years (since the New Economic Policy was introduced in 1970), is how to review these policies to make them more acceptable to all the races, and to enable these affirmative-action policies to ensure that the Malaysian economy becomes even more internationally competitive?

This challenge becomes more relevant with the onset of globalisation.

The NEP has inadvertently also caused a "dependency syndrome" in some quarters. This is why former Prime Minister Dr Mahathir and the Deputy Prime Minister Abdullah Ahmad

Badawi have both urged Malaysians to adopt "a change in mindsets" to reduce this "dependency syndrome" and to become more competitive internally and externally.

Globalisation Challenges

Globalisation undoubtedly poses a great challenge to Malaysia's capacity to be internationally competitive. The challenge now is how to make Malaysia cope more effectively with the powerful and sometimes disruptive forces of globalisation—without compromising the policies of affirmative action and national sovereignty?

The notion that globalisation stands for a borderless world, where sovereign countries will lose their economic and even dilute their political independence, raises much concern and emotional reaction all over the Third World. Malaysia is no exception!

If globalisation is phased in carefully and prudently, without exacerbating these legitimate fears of losing sovereignty to the huge multinationals and foreign countries, then globalisation will be more welcome.

But there is also the legitimate fear that too much globalisation too soon, will undermine the achievements and prospects of the affirmative-action policies. These policies are administered to benefit the disadvantaged Malays and other indigenous races called the *Bumiputeras* as well as all the other poor Malaysians, regardless of race.

The challenge then is to change the mindset of all those Malaysians who have grown up to believe that the government will provide them with privileges for advancement under the affirmative-action policies, regardless of globalisation and keener international competition.

However, it is not an easy task as some aspects of a "dependency syndrome" have become entrenched in the psyche of some sections of the Malaysian society. This syndrome is found mainly in the rural areas and across the education system and even in the recruitment for employment in the public services.

Unfortunately, there are still too many vested groups that want to perpetuate their privileged positions, even at the highest levels of the public service and the business sector.

But it is heartening that former Prime Minister Dr Mahathir and the present Prime Minister, Abdullah Ahmad Badawi, have managed to introduce the concept of meritocracy in the education system, to get Malaysia to become more internationally competitive. But this is proving difficult.

Thus recently the government decided to teach the English language more extensively in schools, especially in Mathematics and Science. This important policy change will undoubtedly enable Malaysian graduates at all levels to become more open minded to competition, liberalisation and to globalisation.

The Medium-Term Economic Outlook

The Malaysian economic outlook in the medium term is however encouraging even at this time of considerable international economic uncertainty. Its economic fundamentals are basically strong. The economic growth has been steady at around 7 per cent per annum for many years. The balance of payments have performed well. The national reserves remain at about 5 months worth of retained imports and the national foreign debt and inflation are internationally recognised to be low.

The only weak fundamental is the Federal Budget. The Budget has been in deficit for about 5 years now as a result of the government's attempts to combat the global slowdown through countercyclical measures.

There is much pressure from the private sector for the government to expand its public expenditures. However, government cannot afford to be more counter cyclical, in view of these mounting Budget deficits.

The challenge would be to reduce these deficits and to bring about balanced budgets as soon as possible.

But it was because of the generally strong economic fundamentals that the economy was able to successfully weather the 1997 Asian financial crisis. The IMF's straitjacket policy recommendations were rejected and instead Malaysia used its own brand of economic strategies that used the unorthodox foreign exchange controls.

These radical policies helped Malaysia combat the unfettered western currency speculators and speeded up economic recovery without having to experience the disruptive social upheavals faced by countries that accepted the IMF's advice.

However, most Western analysts, including the International Monetary Fund (IMF), initially criticised these unconventional measures. But these very measures have proven successful and the IMF has since had to confess that exchange controls are permissible under its rules and that Malaysia was right in introducing them!

The Malaysian economy recovered rapidly after the 1997 Asian financial crisis and was well sustained until the war in Iraq and the Severe Acute Respiratory Syndrome (SARS) broke out and restrained its economic expansion and progress.

The international economic uncertainty has however lessened with the end of the war in Iraq. The progress made in combating the SARS threat which caused low business and consumer confidence has also been sustained.

Thus the Malaysian economy is expected to move onto an upward trend of strong recovery and continuing economic growth.

But the challenge would continue to be—how to further restructure the economy, to sustain a high rate of economic growth and distribution. The economy has been vulnerable because of the heavy dependence on the electronics industry and the recent weak foreign demand for electronics manufactures and components.

The service industries which constitute only about 52 per cent of the GDP will have to be expanded at a much faster rate, if Malaysia is to further diversify its economy and be able to earn higher value-added export earnings.

WTO Challenges

The WTO would confront Malaysia with challenges to liberalise at a faster pace.

In manufacturing, Malaysian industrial tariffs have been relaxed considerably, especially within the Asean Free Trade Area (AFTA) where the average import rates are 0-5 per cent.

However, in the service sector, the list of Offers and Requests could be expanded further.

The professional groups like the architects, bankers, doctors, engineers, lawyers and other professions, tend to be less open to change and prefer to be protected. *The government could take tougher measures to prod the professional groups to liberalise more or at least according to a mutually agreed time schedule, that should cover all the service industries and professions.*

In banking, insurance, shipping, air services, there would be more resistance. This is because, compared to other developing countries, much more liberalisation has taken place in these areas at least so far.

At the same time, there is no concerted international effort to liberalise much faster. No nation (and especially the developing countries) would want its essential service industries to be dominated by foreign interests. Nevertheless, there is plenty of room for the provision of better customer services, which only greater competition can give.

The slowdown in the WTO negotiations may not be entirely unwelcome. It does give more time for the developing countries to consolidate their positions and to prepare to negotiate more effectively with the industrial countries. The rich countries may want to take the lion's share in these unequal WTO negotiations, particularly in those negotiations that are undertaken in the so-called "Green Rooms" that are organised by the rich countries of the WTO!

For Malaysia, with AFTA now in full operation, we have to review how well Malaysia is faring. AFTA was introduced only in January 2003 and so it is difficult to ascertain the full impact on Malaysia's capacity to compete in the 250 million AFTA market of Asean.

One major challenge of AFTA for Malaysia would be, how to give up of the preferential treatment enjoyed by the national car—the Proton Saga, in 2005!

With so little time left it might be more urgent now to form some strategic alliances with regional or international automobile corporations to raise Malaysia's competitiveness and access to the wider international markets.

Even now Malaysia's competitiveness within AFTA could be eroding. This is because it is clear to Malaysian businessmen that if the other nine AFTA countries offer more favourable business and investment opportunities, then even Malaysian businessmen would be persuaded to transfer their operations to other AFTA countries. From there they could still export to the whole AFTA region, including exporting back to Malaysia!

In fact, the forces of globalisation will put more pressure on Malaysia to become more globalised not only in our trade and investment activities, but in the whole range of international diplomatic relations

Revising the Look East Policy

Hence the Look East Policy which Prime Minister Mahathir rightly introduced about 20 years ago may have to be revised.

This policy aimed to change the mindsets of Malaysians who, under the former British colonial regime, were directed to look up mainly to the United Kingdom and the West as the best examples to be followed in all fields.

The Look East Policy on the contrary encouraged Malaysians to look to Japan, and the successful countries in East Asia, like South Korea, China and Taiwan, for inspiration from their achievements and values. For instance, Asians give greater priority to safeguard the societal interests rather than individual rights that might conflict with broader social and cultural values and human welfare.

Discipline, hard work and social cohesion are seen to be stronger values in the East. The example of how Japan as an Asian country, has succeeded in becoming a foremost

modern industrial country, yet preserving its precious
cultural heritage, has been promoted for emulation.

*The Look East Policy has helped to counter the former
colonial psychology that 'white is right' and that the West is Best!*
Malaysians were encouraged to go back to their cultural
roots and seek inspiration from their Asian values and from
the former glory of the Islamic civilisation.

The challenge now is to adjust the Look East Policy, to
shift it more to the centre. This would entail the acceptance
that there is also much that can be learnt from the West and
that we need to be more open to ideas and influences from
both the East and the West.

A Relook at "Malaysia Incorporated"

Malaysia Incorporated policies pose another challenge. This
policy was introduced by Dr Mahathir to overcome the
serious gap in cooperation between the country's public and
private sectors.

It is again a legacy of the colonial past when the British
government kept its civil servants detached and aloof from
local businessmen, but close to their own British business
interests.

Thus even after Malaysia's Independence from Britain in
1957, civil servants like me were discouraged from dealing
with our own businessmen for fear that we might get too
familiar and compromise our official positions with business
interests.

*There was this concern that local businessmen would corrupt
the officials and gain undue favours. At that time civil servants
were brought up to believe that businessmen are greedy and had
no interest in the promotion of national development and public
welfare, except to promote their own selfish interests!* This
attitude carried on even after Independence and was easier

to justify, as the big business was mainly in the hands of foreigners and a few non-Malays.

However, the Malaysia Incorporated policies changed these negative attitudes. Dr Mahathir convinced most civil servants that the more profits the businessmen make, the more taxes could be collected. Then the government could do more to promote socioeconomic development to benefit the people, the civil servants themselves and especially the poorer sections of the community.

Malaysia Incorporated has been successful but the challenge now is to reduce the proximity of the government with some big businesses. But how do we do it?

Some relationships between government and big business could get too cosy. Already there are wide spread perceptions that contracts are negotiated at unduly higher than market prices and that corruption is rising! These unfavourable perceptions must be studied and the Malaysia Incorporated policies duly refined to improve transparency and good corporate governance.

Upgrading of SMIs

The SMIs in Malaysia are as elsewhere the backbone of domestic industrialisation in Malaysia. Presently, it is the large plantations, the national oil company Petronas and the giant utilities and manufacturing enterprises that provide most of the output in the country. But the SMIs provide the support services and the bulk of employment.

Most of the modern export-oriented industries are largely owned by foreign multinationals and the few Malaysian multinationals. Thus it is important to build these local SMIs to grow them into bigger companies which would be the Malaysian multinationals of the future.

The challenge is how to strengthen these SMIs?

Much has been done through the provision of soft loans, skills training and incentives, to increase production and exports. Nevertheless, most of our the SMIs are still relatively slow to expand and to become internationally competitive.

One serious constraint is that many of these SMIs are old Chinese family companies that are conservative and resistant to change. They could be complacent as well as comfortable with their present position of earning small but reliable profits.

The other problem is that many of these SMIs may not have adequate access to the government's financial assistance and technical advice because of language and communication problems with government agencies.

Furthermore, the SMIs with capital of less than RM2.5 million (US$1 = RM3.80), are subject to the government requirement that they have to obtain permission before they can expand their investment. Besides the government's intention to monitor the development of these SMIs, another important aim is to encourage these SMIs to provide about 30 per cent of their new investment, for *Bumiputera* ownership, at prices that may not be attractive to the owners.

In addition there is the disincentive for these SMIs that their long standing family ownership of these SMIs would be diluted by outsiders who may not have the same commitment to their traditional family business practices.

The challenge therefore is for the government to consult more closely with these largely Chinese-owned SMIs, to find out the right formula,that would encourage them to expand and to welcome more *Bumiputera* participation,in accordance with the country's national affirmative-action policies.

Perhaps some form of incentives for these family-owned SMIs would help encourage greater cooperation between the essentially non-Malay SMIs and their potential *Bumiputera* partners?

The China Challenge

China is increasingly providing tough challenges and also vast opportunities to the SMIs and to the Malaysian economy. With China's huge 1.2 billion consumer market, low wages, and its determination to become more technological and productive, our SMIs will find it very difficult to compete with China. The labour-intensive SMIs in Malaysia will have less competitive advantages compared to their counterparts in China.

Many Malaysian businessmen will want to follow the trend set by the multinationals that are moving to China, since China recently joined the WTO.

Hence, Malaysian SMIs will need to find new and higher value added products as well as new markets to compete with China.

Although there are also opportunities in China's vast market, there are no significant inroads that have been made by Malaysians so far. Hence the question arises as to whether the government and local SMIs will be able to combine their forces, to raise their competitive capacity to face the challenges from China? As of now the signs are not encouraging.

Another challenge from China is its rising capacity to attract more FDIs from other parts of the world and away from Asean countries, including Malaysia.

With all this severe international competition for FDIs, Malaysia will have to face up to the challenge of making it more attractive for FDIs to flow into Malaysia.

Malaysia's foreign ownership policies have to be further liberalised to compete with the more relaxed policies in competing countries. The Foreign Investment Committee that deals with foreign ownership has to be further revamped, or its policies should be made less stringent for FDIs.

Alternatively, because FDIs are becoming harder to come by, the government is also currently looking into more ways and means of increasing domestic consumption and investment.

Thus Dato' Seri Abdullah while he was the Acting Prime Minister, announced the need to give higher priority to establish "agro-based industries" in Malaysia. The challenge is therefore to attract more FDIs and/or to compensate for its decline, by aggressively promoting greater domestic investment. But we should aim to achieve both objectives, and not pursue just one or the other goal.

Weak Research & Development (R&D)

Another challenge is to increase Research and Development (R&D) to play a vital role in raising our competitiveness and to face up more confidently to globalisation. But our record of research and development has not been impressive.

R&D arguably constitutes only about 0.5 per cent of the Malaysian GNP, depending on what definitions are used. This is a far cry from what is required to increase value-added production and productivity in the economy. *Because of this low investment in R&D, Malaysia's Incremental Capital Output Ratio (ICOR) is not favourable*. It is about 4:1, meaning that we have to spend RM4 to earn RM1 worth of output. In other words, Malaysian investment is comparatively less efficient. This could be due to the high investment in social infrastructure such as for schools, hospitals and roads and railways, for which the economic returns are low and slow. It could also be due to inefficient spending.

Political Challenges Increasing

Malaysia has enjoyed enviable political stability for 47 years, ever since Independence!

Malaysia has also had the same Coalition political party called the Barisan Nasional or National Front all this time. Although there is a very active and vociferous Opposition, the government nevertheless has ruled with a large majority of over two-thirds in the Federal Parliament.

This fact alone has given the country the much-needed peace and stability to enable impressive economic development and to attract foreign investment, business and tourists.

The Opposition comprises two major political parties: PAS, the Islamic party, and the Democratic Action Party (DAP), besides many other smaller but active political entities.

The challenge for the government is to maintain its strong Parliamentary majority of two-thirds. It has become increasingly difficult to achieve this objective due to the competition from the Islamic Party (PAS), which seeks more Islamisation of the country.

The DAP is mainly supported by the Chinese voters who want more benefits for the strong Chinese population (27 per cent) in the country.

The Barisan Nasional coalition party is composed of major political parties such as the United Malays National Organisation (UMNO), the Malaysian Chinese Association (MCA), Parti Gerakan, the Malaysian Indian Congress (MIC) and several other smaller political parties.

These government political parties now enjoy the support from the majority of the Malays, the Chinese the Indians and the other smaller minority ethnic groups, like the Ibans and Dayaks and others from Sabah and Sarawak.

The challenge for the ruling Barisan Nasional is to seriously examine the causes of the expanding support for the DAP and the PAS and to seek ways to overcome the grievances raised by those who oppose the governing Barisan party.

Some of the apparent reasons for voters going over to PAS could ironically be due to the government's rapid socioeconomic progress achieved in so short a period since Independence and particularly in the last 20 years.

Modernisation in the lifestyles of the new generation of educated and liberal Malays and those who have lived abroad could go against the grain of social and religious conservatism, particularly in the rural areas of the country.

In any case, the resurgence of Islam in Malaysia and in many other countries could also create reactions to what are sometimes regarded as imported and undesirable foreign influences.

The growing gap between the rich and the poor, as in any rapidly developing society, could also be providing fertile ground for dissatisfaction and opposition to the ruling political parties. The often ostentatious lifestyle of the elitist groups are presented by the opposition parties and to the conservative religious groups, as signs of decadence and corruption which should not be tolerated.

But, more importantly, there could be deep religious convictions that drive some very conservative religious groups to resist politics that are viewed to be too liberal and tolerant. This is the brand of religion of whatever faith, that although in the minority, could nevertheless support extremism and terrorism and pose security problems anywhere in the world.

The challenge therefore is to continue to strengthen religious freedom in Malaysia, but to be vigilant against extremism of all kinds that could threaten national security.

Revising the Internal Security Act (ISA) 1960

Hence the government with the backing of the majority of the voters has maintained the need for the Internal Security Act (ISA) 1960 which has helped to secure peace and stability in the country.

The introduction of similar laws of preventive detention in the U.S. and other industrial countries (which had condemned the Malaysian government before the September 11 terrorist attacks in the U.S.), has strengthened the Malaysian government's position on this sensitive issue.

However, the Human Rights Commission (Suhakam) that was established by the government about three years ago has recently recommended the repeal of the ISA and for its replacement with a new comprehensive legislation, that has better safeguards against the infringement of human rights.

The challenge then is to strike the right balance between ensuring national security and protecting the citizens' human rights.

Malaysia: The Modern Islamic State

Islamisation has gained much ground as the Islamic political party PAS has won more support and increased its political influence in recent years.

This could be due to the rising resurgence world wide of the recognition for a greater role for religion and spirituality to attain world peace and stability.

Islam is the official religion of Malaysia but Federal the Constitution provides for freedom of religion for all faiths.

The Malays who constitute about 60 per cent of the total population of about 24 million are practically all Muslims. They are tolerant and peace loving. Nevertheless, as in all religions, there is a narrow extremist fringe that the government is wary about. Many of the more extreme Islamic groups have drawn their inspiration from some educational institutions in the Middle East.

Furthermore, there is a large number of students who attend private Islamic religious schools in the country. On graduation therefore they find it more difficult to get suitable employment because of the mismatch between their religious qualifications and job opportunities. Thus they are a source of dissatisfaction and frustration and this can have some negative implications on social stability.

The challenge for the government is therefore to reduce the influence of extremist groups. This challenge is now being met by bringing these religious schools into the mainstream educational system, where more emphasis is given to science and technology. This move will enable better employment opportunities for the graduates of religious schools.

But the government has also been very firm in monitoring and combating extremism of all kinds. There is thus close regional cooperation within Asean countries and with other countries that are committed to fighting international terrorism.

Improving International Relations

Malaysia foreign policy is based on being friendly with all countries without taking sides or joining any blocs, even before the Cold War.

It is a strong supporter of the United Nations and has in fact taken part in many U.N. Peacekeeping Missions all over the world.

Malaysia takes a leading role as an active member of the Non-Aligned Movement (NAM) and the Organisation of Islamic Conference (OIC). Malaysia is currently the Chair for both these international organisations and Prime Minister attended the recent G8 Meeting in Evian, France, because of his chairmanship of NAM and the OIC.

Indeed Malaysia can for these very reasons and its high standing in the Islamic as well as the Western world, contribute significantly in enhancing the prospects of the new U.S.-brokered road map to peace and prosperity in the Middle East.

It is fair to say that Malaysia is now widely regarded as a model of a progressive, liberal and progressive, democratic Islamic country which is also seen as an advanced developing country.

The challenge for Malaysia is therefore to promote itself as a model country, not only for all developing countries but also for all Islamic countries.

Unfortunately, however, the U.S. and several other Western countries often overreact to Malaysia's steadfast criticism of some of their policies of discrimination and oppression in international finance, trade, politics and security policies.

Malaysia has consistently fought for equity and justice and a fairer deal for all developing countries and resisted hegemony that is sought by a few powerful and rich industrial countries. But this stance has upset some rich countries.

Malaysia has been able to take on this leadership role on behalf of Third World countries because it is not obliged to the rich countries for aid or debt relief.

Indeed there are many developing countries that are grateful to Malaysia for representing their views at international fora like the U.N., the WTO, the World Bank and the IMF, all of which are perceived to be dominated by some rich and powerful Western countries.

Prime Minister Mahathir had actively promoted his philosophy of "Smart or Strategic Partnership", which aims to promote "prosper-thy-neighbour" as opposed to the "beggar-thy-neighbour" policies of some industrialised countries.

All these different approaches adopted by Malaysia in its international relations, have won much support from the Third World but unfortunately cause a few problems with some powerful western countries that resent Malaysia's independent stance.

Perhaps the challenge is for Malaysia or Dr Mahathir to modify some of its tough international positions, so as to be regarded as more 'friendly' and acceptable to some of these hegemonistic countries.

Because Malaysia's international presentation has been so robust and sometimes quite blunt, Malaysia's views as expressed by its leaders, sometimes sound unbearably harsh to a few powerful Western countries. They have not been used to hearing hard talk from developing countries, particularly their former colonies!

Many Third World countries often dare not take the risk of being even justifiably critical, as they fear reprisals in the form of the withdrawal of aid and debt relief and the reduction in security protection. Worse still, now we have the new threat of the so-called "pre-emptive attack"!

The challenge for some powerful countries in the Western World is to appreciate the rapid political and economic development and progress made by Malaysia as a developing and a modern Islamic country. The need to work together with Malaysia to promote the country as a model for other developing and Islamic countries to mutually benefit from our experiences in good government!

Malaysia at the Crossroads

Malaysia is at an important crossroads. The transition from Dato' Seri Dr Mahathir Mohamad to Dato' Seri Abdullah Ahmad Badawi as the next Prime Minister from November 2003 is a major watershed in Malaysia's history and its future!

Dr Mahathir has groomed his successor Abdullah very carefully and thus the transfer of authority was smooth and stable. Indeed, on his return to Malaysia in April 2003 after a two-month holiday, Dr Mahathir commended Abdullah for covering his duties as Acting Prime Minister "with flying colours"! Malaysia's socioeconomic, political and foreign policies are also anticipated to continue, with Abdullah adopting much of Dr Mahathir's policies that have served the nation so well.

Malaysia as a united and modern democratic Islamic country with a free enterprise economic system, therefore has the capacity and confidence to forge ahead on its path to become a developed nation by 2020, according to Malaysia's own distinctive Vision 2020 national goals and aspirations.

With the goodwill and support of the international community and the determination, dedication and drive of our leaders and our people, Malaysia will continue to prosper and progress with peace and stability. Indeed Malaysia could set the example for the Third World as it moves towards First World status. There is much confidence that Malaysia will be able to achieve the high aspirations that it has set for itself.

The contribution of the new Prime Minister, Dato' Seri Abdullah Ahmad Badawi, will therefore be crucial to Malaysia's march towards the aspirations of Vision 2020.

Suggested Reading

Abraham, Collin, *The Naked Social Order: The Roots of Racial Polarisation in Malaysia*, Subang Jaya: Pelanduk Publications, 2004

Bell, Daniel, *The Future of Technology*, Subang Jaya: Pelanduk Publications, 2001

Blustein, Paul, *The Chastening: Inside the Crisis that Rocked the Global Financial System and Humbled the IMF*, New York: Public Affairs, 2001

Chandler, Clay, "Coping with China," *Fortune*, Vol. 147, No. 1, 2003

Chang, Gordon G., *The Coming Collapse of China*, London: Century, 2002

Clifford, Mark L., and Pete Engardio, *Meltdown: Boom, Bust, and Beyond*, Paramus, New Jersey: Prentice Hall Press, 2000

Frank, Thomas, *One Market Under God: Extreme Capitalism, Market Populism, and the End of Economic Democracy*, New York: Doubleday, 2000

Friedman, Thomas, *The Lexus and the Olive Tree*, London: HarperCollins, 2000

Giddens, Anthony, *Runaway World: How Globalisation is Reshaping Our Lives*, London: Routledge, 2000

Gilpin, Robert, *The Challenge of Global Capitalism: The World Economy in the 21st Century*, Princeton: Princeton University Press, 2000

Goldstein, Morris, Graciela Kaminsky and Carmen Reinhart, *Assessing Financial Vulnerability: An Early Warning System for Emerging Markets*, Washington, D.C.: Institute for International Economics, 2000

Gomez, Edmund Terence, *Malaysia's Political Economy: Politics, Patronage and Profits*, Cambridge: Cambridge University Press, 1999

James, Harold, *The End of Globalisation: Lessons from the Great Depression*, Cambridge: Harvard University Press, 2001

Jomo K.S., *Malaysian Eclipse: Economic Crisis and Recovery*, London: Zed Books, 2001

Jomo K.S. (ed.), *Southeast Asia's Industrialization: Industrial Policy, Capabilities and Sustainability*, Basingstoke, Hampshire: Palgrave Macmillan, 2001

Jomo K.S. and Mushtaq Khan (eds.), *Rents, Rent-Seeking and Economic Development: Theory and the Asian Evidence*, Cambridge: Cambridge University Press, 2000

Jomo K.S. and Ng Suew Kiat, *Malaysia's Economic Development: Policy and Reform*, Subang Jaya: Pelanduk Publications, 1996

Jomo K.S. and Shyamala Nagaraj (eds.), *Globalization Versus Development: Heterodox Perspectives*, Basingstoke, Hampshire: Palgrave Macmillan, 2001

Kepel, Gilles, *Jihad: The Trail of Political Islam*, translated by Anthony F. Roberts, Cambridge, Mass.: The Belknap Press/Harvard University Press, 2002

Kristof, Nicholas D., and Sheryl WuDunn, *Thunder from the East: Portrait of a Rising Asia*, New York: Alfred A. Knopf, 2001

Lardy, Nicholas R., *Integrating China into the Global Economy*, Washington, D.C.: Brookings Institution Press, 2002

Leadbeater, Charles, *Up the Down Escalator: Why the Global Pessimists Are Wrong*, London: Viking, 2002

Legrain, Philippe, *Open World: The Truth About Globalisation*, London: Abacus, 2002

Lewis, Bernard, *What Went Wrong? Western Impact and Middle Eastern Response*, Oxford: Oxford University Press, 2002

Lindsey, Brink, *Against the Dead Hand: The Uncertain Struggle for Global Capitalism*, London: John Wiley, 2002

Lowenstein, Roger, *When Genius Failed: The Rise and Fall of Long-Term Capital Management*, New York: Random House, 2000

Mahathir Mohamad, *Mahathir Mohamad: A Visionary and His Vision of Malaysia's K-Economy*, Subang Jaya: Pelanduk Publications, 2002

Mahathir Mohamad, *Reflections on Asia*, Subang Jaya: Pelanduk Publications, 2002

Mahathir Mohamad, *Globalisation and the New Realities*, Subang Jaya: Pelanduk Publications, 2002

Mahathir Mohamad, *The Malaysian Currency Crisis: How and Why It Happened*, Subang Jaya: Pelanduk Publications, 2000

Mahathir Mohamad, *A New Deal for Asia*, Subang Jaya: Pelanduk Publications, 1999

Mahathir Mohamad, *The Challenges of Turmoil*, Subang Jaya: Pelanduk Publications, 1998

Mahtaney, Piya, *The Economic Con-Game: Development: Fact or Fiction?*, Subang Jaya: Pelanduk Publications, 2002

Mayer, Martin, *The Fed: The Inside Story of How the World's Most Powerful Financial Institution Drives the Markets*, New York: The Free Press, 2001

McQueen, Humphrey, *The Essence of Capitalism*, London: Profile Books, 2001

Micklethwait, John, and Adrian Wooldridge, *A Future Perfect: The Challenge and Hidden Promise of Globalization*, New York: Crown Business, 2000

Mills, John, *Managing the World Economy*, London: Macmillan, 2000

Muzaffar, Chandra, *Rights, Religion and Reform: Enhancing Human Dignity Through Spiritual and Moral Transformation*, London: RoutledgeCurzon, 2002

Navaratnam, Ramon V., *Malaysia's Economic Sustainability: Confronting New Challenges Amidst Global Realities*, Subang Jaya: Pelanduk, 2002

Navaratnam, Ramon V., *Malaysia's Economic Recovery: Policy Reforms for Economic Sustainability*, Subang Jaya: Pelanduk, 2000

Navaratnam, Ramon V., *Healing the Wounded Tiger: How the Turmoil is Reshaping Malaysia*, Subang Jaya: Pelanduk, 1999

Navaratnam, Ramon V., *Strengthening the Malaysian Economy: Policy Changes and Reforms*, Subang Jaya: Pelanduk, 1998

Navaratnam, Ramon V., *Managing the Malaysian Economy: Challenges and Prospects*, Subang Jaya: Pelanduk, 1997

Okposin, Samuel Bassey, and Cheng Ming Yu, *Economic Crises in Malaysia: Causes, Implications and Policy Prescriptions*, London: Asean Academic Press, 2000

Panitchpakdi, Supachai, and Mark L. Clifford, *China and the WTO: Changing China, Changing World Trade*, Singapore: John Wiley & Sons (Asia), 2001

Pempel, T.J. (ed.), *The Politics of the Asian Economic Crisis*, Ithaca, New York: Cornell University Press, 1999

Purves, Bill, *China on the Lam*, Hong Kong: Asia 2000, 2002

Pillar, Paul R., *Terrorism and U.S. Foreign Policy*, Washington, D.C.: Brookings Institution Press, 2001

Rashid, Ahmed, *Jihad: The Rise of Militant Islam in Central Asia*, New Haven: Yale University Press, 2001

Rashid, Ahmed, *Taliban: Militant Islam, Oil and Fundamentalism in Central Asia*, New Haven: Yale University Press, 2000

Rosenberg, Tina, "The Free-Trade Fix," *The New York Times Magazine*, August 18, 2002

Singer, Peter, *One World: The Ethics of Globalization*, New Haven: Yale University Press, 2002

Soros, George, *On Globalisation*, New York: Public Affairs, 2002

Soto, Hernando de, *The Mystery of Capital: Why Capitalism Triumphs in the West and Fails Everywhere Else*, New York: Basic Books, 2001

Stiglitz, Joseph E., *Globalisation and Its Discontents*, London: Allen Lane/The Penguin Press, 2002

Stiglitz, Joseph E., "A Fair Deal for the World," *The New York Review of Books*, May 23, 2002

Stiglitz, Joseph E., "The Insider: What I Learned at the World Economic Crisis," *The New Republic Online*, April 17, 2000

Studwell, Joe, *The China Dream: The Elusive Quest for the Greatest Untapped Market on Earth*, London: Profile Books, 2002

Talbott, Strobe, and Nayan Chanda (eds.), "Empowered through Violence: The Reinvention of Islamic Extremism," in *The Age of Terror*, The Perseus Press, 2001

Tariq Ali, *The Clash of Fundamentalisms: Crusades, Jihads and Modernity*, London: Verso, 2002

Vines, Stephen, *The Years of Living Dangerously: Asia—From Financial Crisis to the New Millennium*, London: Orion Business Books, 2000

Vogel, Ezra F., *Is Japan Still Number One?*, Subang Jaya: Pelanduk Publications, 2000

Volpi, Vittorio, *Japan Must Swim or Sink*, Subang Jaya: Pelanduk Publications, 2001

Ye Lin-Sheng, *The Chinese Dilemma*, Sydney: East-West Publishers, 2004

Zachary, G. Pascal, *The Global Me: New Cosmopolitans and the Competitive Edge: Picking Globalism's Winners and Losers*, New York: Public Affairs, 2000

INDEX

Abdul Aziz Shamsuddin, 91, 243
Abdul Razak Hussein, 196, 219, 232
Abdullah Ahmad, 77
Affirmative-action policies, 2, 4, 34, 76, 189, 268-269, 277
AFTA, see Asean Free Trade Agreement
Agilent Technologies, 118
Ahmad Hassan, 91
Ahmad Husni Hanadzlah, 92
Ahmad Talib, 129
AISP, see Asean Integration System of Preferences
Alliot-Marie, Michele, 167
Al-Qaeda, 176
AMCHAM, see American Malaysian Chamber of Commerce

American Malaysian Chamber of Commerce, 96-97
Arnett, Peter, 81
Asean, see Association of Southeast Asian Nations
Asean Free Trade Agreement (Asean Free Trade Area), 5, 8-9, 18, 30-31, 74-75, 85-86, 114, 146, 172, 272-274
Association of Southeast Asian Nations, 4-6, 8-10, 35-36, 38-39, 74, 85, 111-115, 119, 124, 219, 235, 260, 272-273, 278, 283

Bank Negara Malaysia (Bank Negara), 62, 66-67, 101-102, 125, 130, 137, 209-210
Blair, Tony, 48, 211

REHDA Malaysia, *see* Real
Estate and Housing
Developers' Association
Malaysia
Rice, Condoleezza, 170,
174-175
Rosnani Ibrahim, 143
Royal Police Commission, 225,
238
Rukunegara, 216

S. Jeyakumar, 74, 116
S. Samy Vellu, 20
Saddam Hussein, 49, 83, 85
SAID, *see* Southern Africa
International Dialogue
Samsudin Osman, 32, 170, 197
SARS, *see* Severe Acute
Respiratory Syndrome
Severe Acute Respiratory
Syndrome, 94-96, 98-101,
129-130, 271-272
Shafie Mohd Salleh, 245, 262
Shimizu, Kentaro, 86
Shinawatra, Thaksin, 124
Shiokawa, Masajuro, 29
Singapore-U.S. Free Trade
Agreement, 111
Small and Medium Industries
Development Corporation, 7
SMIDEC, *see* Small and
Medium Industries
Development Corporation
Southern Africa International
Dialogue, 165, 173-174
Stern, Nick, 89
Suhakam, *see Suruhanjaya Hak
Asasi Manusia* (Human
Rights Commission)

Suruhanjaya Hak Asasi Manusia
(Human Rights
Commission), 216, 282

Talib Jamal, 156
Tan Siew Sin, 144
Tan Siok Choo, 128
Technology Investment Fund,
134
Terrorism, 41-42, 49-51, 56-58,
60, 64-65, 68, 89-90, 113,
144-145, 152, 163-165,
167-170, 176, 220, 235, 243,
248, 282-283
THORA, *see* Tun Hussein
Onn Renewal Awards
TIFA, *see* Trade Investment
Framework Agreement
Trade Investment Framework
Agreement, 173, 235
Tun Hussein Onn Renewal
Awards, 12

UMNO, *see* United Malays
National Organisation
United Malays National
Organisation, 14, 146-147,
149, 280

Vision 2020, 21, 39, 75, 77-79,
107, 133, 189, 216, 230, 237,
266, 286
Voluntary Separation
Schemes, 251
VSS, *see* Voluntary Separation
Schemes

Wieczorek-Zeul, Heidemarie,
90

What
MANAGING THE MALAYSIAN ECONOMY:
CHALLENGES AND PROSPECTS (1997)
is all about:

This book discusses the challenges and prospects that lie ahead as Malaysia marches towards the new millennium and beyond. The rapid economic growth Malaysia enjoyed since the mid-1980s is a fact that is hard to ignore. The economy had averaged 8 per cent per annum for the last 8 years. Economists have expressed concern about inflation, but at 3-4 per cent per annum, it is still well under control.

This book encourages a critical and distinctively Malaysian approach to the problems the country faces as it strives towards developed-nation status by the year 2020. Malaysia's plan was to go for productivity-driven growth with sustainable external balance and price stability. Despite Malaysia's inherent problems of a labour force that is becoming increasingly scarce and costly, Malaysia will continue to be attractive to foreign investors in the years to come because of its political stability, good infrastructure and an English-speaking workforce. Malaysia's continued success will now depend on its ability to attract technologically advanced industries and continuing strong leadership.

ISBN 967-978-581-5

What the reviews say of
MANAGING THE MALAYSIAN ECONOMY:
CHALLENGES AND PROSPECTS (1997)

"... a comprehensive and well-written book. The lessons of experience will be useful to us here at the World Bank as well as to policymakers in other developing countries .... I agree with the key challenges on development that [Navaratnam has] identified ..., including the importance of sound economic management, strengthening productivity-driven growth, and addressing skill shortages, environmental decline, and corruption."
James D. Wolfensohn *World Bank President*

"... a thought-provoking book. ... Navaratnam looks at the challenges facing Malaysia squarely and to his credit has come forward with many practical solutions to some of the excesses and hangovers from the prolonged economic boom." *The Star*

"... a book worth reading. ... this book ... stimulates public discussion of Malaysia's major economic issues and the associated challenges and prospects. It ... offers criticisms and suggestions that reflect the concerns of a moderate, independent, pragmatic and visionary Malaysian nationalist who wishes to see the continued growth and development of the country." *New Straits Times*

"... a must-read for the business visitor to the country who could use a clear, insightful summation of why Malaysia is likely to be the country that will prosper longest and best amidst Southeast Asia's ascendancy. ... might be good for investors, but it is even better for Malaysians." *Malaysian Business*

"... reviews the diversity of challenges and prospects facing the Malaysian economy as it approaches the second millennium. The binding metaphor of this collection of pieces is management of the economy of Malaysia as it traverses the watershed to become an industrialised nation." *Management*

"Though not everyone may agree with Navaratnam's stance on the environment, the average businessman would probably concur with the need to advocate productivity-driven growth. ... His book is retrospective, emphasising the strengths in Malaysia's economic planning since *Merdeka*." *Malaysian Industry*

What
STRENGTHENING THE MALAYSIAN ECONOMY:
POLICY CHANGES AND REFORMS (1998)
is all about:

This book attempts to provide solutions to Malaysia's economic malaise as it strives to become an industrialised nation. The Malaysian economy, which had enjoyed spectacular growth for eight consecutive years, was jolted from its euphoria in July 1997, when the vicious contagion effect arising from the *de facto* devaluation of the Thai baht send East Asian currencies and stock markets nosediving to lows the likes of which had never been seen before. Why was this breathtaking march to prosperity brought to a grinding halt by the financial crisis? What were the causes of the crisis and what lessons can be drawn from it?

Though Malaysia has turned the corner in grappling with the financial crisis, tough times still lie ahead. Reforms will take time to be implemented and results will not be forthcoming. A change in mindset is crucial: the crisis must be seen as an opportunity to work towards a more productivity-driven economy. A commitment to reforms and an understanding of how they are to be implemented are vital in expediting recovery and sustaining economic growth. Despite the uncertainty over how long the downturn will last, Malaysia's long-term prospects are still encouraging.

ISBN 967-978-642-0

What the reviews say of
STRENGTHENING THE MALAYSIAN ECONOMY:
POLICY CHANGES AND REFORMS (1998)

"... essential reading for the layman, economists, researchers and policymakers interested to have a quick grasp of how and why the Malaysian economy ticks, from boom to recession. ... [Navaratnam] speaks with refreshing candour and sometimes in perplexed exasperation on the kaleidoscopic issues and challenges confronting the economy. ..." *The Star*

"... a comprehensive assessment of the problems facing the Malaysian economy, ranging from social and economic issues to unethical professional practices, money politics, corruption and nepotism. ... essential reading for the man in the street, policymakers and economists who desire a good understanding of the state of play and of the future outlook for the economy. Our current position at the recessionary phase of the business cycle makes Navaratnam's contribution both pertinent and topical. These essays are ... well pitched to stimulate public policy discussion and debate. ... [He] espouses a common-sense approach to economic policy. His approach is one of steering, gently nudging and carefully explaining. ... his arguments appear cogent and sensible." **Wong Koi Nyen** *Lecturer in Economics and Robin Pollard Head of School of Business and Information Technology, Monash University Sunway, Malaysia*

What
HEALING THE WOUNDED TIGER:
HOW THE TURMOIL IS RESHAPING MALAYSIA (1999)
is all about:

What began in July 1997 as a Thai currency crisis saw the rapid depreciation of Asian currencies and the collapse of stock markets across the region. The Malaysian economy, which had enjoyed spectacular growth for eight consecutive years, was not spared. By introducing capital and currency controls to shield its battered economy from external volatility and currency manipulation, Malaysia took a step away from orthodoxy. However, it has not insulated itself from the mainstream of global economy and will continue to promote trade and foreign direct investment as conduits for universal prosperity. More reforms in the financial system are still needed to counter the rising tide of globalisation. There is more scope for financial sector liberalisation to allow for both the infusion of capital and expertise into the system, which will contribute to long-term stability. While absolute liberalisation in the system may not be possible in the near term, greater relaxation of rules would pave the way to recovery.

ISBN 967 978 674 9

What
MALAYSIA'S ECONOMIC RECOVERY: POLICY REFORMS FOR ECONOMIC SUSTAINABILITY (2000)
is all about:

As East Asia picks up the pieces of its burst bubble, new questions emerge: What have we learnt from the crisis? How do we go about ensuring that we will be able to better face the challenges of globalisation in the future? How can companies, especially the smaller ones, prepare for changes and not suffer a repeat of the crisis? How should be regard the rich and powerful industrial countries? Have we become too complacent for comfort? Have we got the will? Are we still on track towards Vision 2020?

Once again, Ramon V. Navaratnam shows us the state of the Malaysian economy and what the government and companies need to do to build a sustainable economy and reshape Malaysia's vision in the 21st century. Navaratnam argues that we must learn from the lessons of history and change for the better. With greater determination and stronger national unity, Malaysia will be able to become more competitive and continue to manage the risks and challenges of globalisation and prosper. He offers pointers on how the Malaysian government and companies can build upon the damage brought about by the Asian financial crisis of 1997. He believes that Malaysia has to undertake more reforms at all levels of the economy to overcome the challenges from within and outside the country, especially in the new era of globalisation.

ISBN 967-978-736-2

What the reviews say of
MALAYSIA'S ECONOMIC RECOVERY: POLICY REFORMS FOR ECONOMIC SUSTAINABILITY (2000)

"... makes interesting reading. ... a wealth of information. ... [With this book, Navaratnam] has contributed to the building of Malaysia's institutional memory." **Dato' Seri Abdullah Ahmad Badawi** *Deputy Prime Minister, Malaysia*

"... [Navaratnam] has not only examined the state of Malaysia's economy, but has also constructively outlined his thoughts on building a sustainable economy and reshaping Malaysia's vision in the 21st century." **Tan Sri Jeffrey Cheah** *Sunway Group*

"A must-read for those interested in what Navaratnam calls 'Malaysianomics'—Malaysia's special brand of pragmatic and innovative approach to economic management. This book brings a fresh view on the Malaysian economy after the crisis. he does not hesitate to defend Malaysia against unfair foreign criticisms but at the same time is concerned enough to suggest ways to improve the economy." **Professor Dr Mahani Zainal Abidin** *Head, Special Consultancy Team on Globalisation, National Economic Action Council, Economic Planning Unit*

"What I like most about the book is the frankness with which the author refers to weaknesses in our institutions and recommends overdue reforms in the economic and banking sectors and in Malaysian public administration. ... 'Crisis concentrates the mind,' it has been said, and this book, by one who has certainly concentrated his mind, deserves to be read not only by students of economics and business executives but also by citizens who are concerned with public issues." ***The Sun***

"... seeks to fathom the depths of the 1997 financial crisis. [Navaratnam] draws lessons from it and sees ways to reshape the Malaysian economy in the face of globalisation. ... also urges reforms and suggests how the government and private sector can benefit from the challenges ahead." ***Going Places***

"It is a commendable and important work, one that, in plain prose, outlines basic problems and offers solutions candidly—sometimes boldly. ... The book rides the waves of popular consensus, yet also urges the debate forward on various issues. It is not the first, nor is it the last. It is written in the midst of change. Above all, its writing is compelled by a desire for reform." *The Star*

"Malaysia's economic development process and experience is unique. Sadly, though, there is scant literature on it. ... The exception, perhaps, is Navaratnam, who has not only ... tracked the country's economic policies, their implementation, impact and effectiveness but at the same time provided the badly needed economic literature on the Malaysian experience." *New Sunday Times*

"This book provides a highly readable account of the Malaysian economic and business situation in 2001. Navaratnam writes with clarity and does not hesitate to express forceful, sometimes controversial views on the issues which contributed to the Asian financial crisis of 1997. He discusses the socioeconomic policies and management of the Malaysian economy since that time and he then calls on the Malaysian government and on global corporations to initiate new thinking and innovative policies in order to build economic sustainability. In some areas his policy prescriptions are primarily relevant to Malaysia (for example, he has interesting things to say about the need to revisit current banking and *Bumiputera* partnership policies). However, his analysis often focuses on the impact of globalisation on developing economies, and in this respect the book is highly relevant to many topical economic and political debates. This is a book that must be read by anyone interested in the state of the Malaysian economy and associated Malaysian domestic policy. It will also interest anyone who wishes to use the Malaysian experience as a case study of the impact of globalisation on non-western economies. Written as it is by a Harvard-trained Malaysian economist ..., this book can provide a perspective from within this young economy, to counteract an overreliance on analyses from outside. **Professor Gill Palmer** *Dean, Faculty of Business and Economics, Monash University, Australia*

"This book presents a different but excellent perspective to managing economic developments in Malaysia, amid globalisation that is rapidly putting more pressure on countries to reform. Navaratnam presents more than a fresh point of view." **Business Today**

"Navaratnam's book has the merit of combining an analytical approach to research with plain, communicative language. Published when the Malaysian economy was beginning to recover from the Asian financial crisis of 1997, it offers a timely and cogent analysis of what Malaysia needs today to build a better tomorrow. And yet despite the encouraging trends, he warns that Malaysians cannot afford to be too confident. Besides describing the opportunities, he also offers insightful advice on the major difficulties Malaysia may encounter in the years ahead by highlighting the importance of policy reforms and the potential perils if these are not done. 'I believe that with the right will and strong national unity, we will win the battle, despite the savage process of globalisation. But to gain from globalisation, we must overcome our own dilemmas,' he advises.

"The subject of Navaratnam's pragmatic investigation is very much part and parcel of the problems the country has to face and solve in the next millennium in achieving the targets of Vision 2020. Malaysia's multicultural milieu, national unity in diversity, productivity, and smart partnership between the private and public sectors are all concepts far too precious to be ignored, and are among the leitmotiv of his farsighted perspective. Though you may not agree with all of his provocative, practical and far-reaching solutions, those interested will find food for healthy discussion and debate.

"... this book will bring about greater Malaysian participation in the making of a developed Malaysia in an increasingly globalised world. The concerned Malaysian reader and 'all those who believe that we can in our own small way contribute to steer the Malaysian economy steadily forward' should listen, change their attitude and thus improve their ability and willingness to contribute to the country's future." **Dr Marie-Aimée Tourres** *Economist, Institute of Strategic and International Studies (ISIS), Malaysia*

"... an excellent synopsis of Malaysia's domestic policy and the nation's position in the global economic and political community. It provides a comprehensive commentary on how Malaysia sees herself and how she perceives others see her. It is essential reading for all those interested in the range of contemporary economic and political issues facing this rapidly growing and changing nation. It is a brief but valuable overview for Malaysians, but also for all those interested in understanding this dynamic nation.

"The text takes the reader brilliantly through the processes of adjustment that are ongoing in Malaysia by intertwining the economic, political and social adjustments that have occurred since the nation's independence and journey through rapid development as one of the major economies of the region.

"The text is more than a description of change but is an analysis of those changes and provides considered views on alternatives for the future.

"It is particularly valuable reading for non-Malaysians, as it is an easy-to-read treatise of a complex and rapidly changing society. The sections on Malaysia's role in the global political and economic environments is particularly valuable." **Lionel Phelps** *Chancellor, Southern Cross University, Lismore, New South Wales, Australia*

What
MALAYSIA'S ECONOMIC SUSTAINABILITY: CONFRONTING NEW CHALLENGES AMIDST GLOBAL REALITIES (2002)
is all about:

The remarkable fact about the Malaysian economy is not how far it has come, but how well it has recovered from the Asian financial crisis of 1997. Tough times now beckon with a slowdown in the Western economies, where the chill winds of global downturn are beginning to nip at the ankles.

Malaysia has been acknowledged as the richest and best managed Islamic democracy in the world. The question, however, is whether Malaysia can sustain this outstanding record of socioeconomic development, with its enviable political stability and its unique brand of social engineering that has strengthened racial harmony in one of the most complex societies in the world. Will Malaysia be able to seize the opportunities that well managed globalisation can provide or will it be "gobble-ised" and dominated by the economic might of the Western industrial powers and the powerful multinationals in the painful process of economic liberalisation.

Malaysian Prime Minister Dato' Seri Dr Mahathir Mohamad has undoubtedly led Malaysia impressively into the 21st century. With him at the helm steering the nation forward, Malaysia is today a modern nation and is moving steadily towards industrialised-nation status by the year 2020. Dr Mahathir, with his deep convictions and political battles in his twenty remarkable years as Prime Minister of Malaysia.

Besides providing an overview of the Malaysian economy and what makes it tick, this book looks at the continuing concerns as to whether Malaysia can overcome the many new challenges that it will confront with the rising tide of globalisation.

ISBN 967-978-804-0

What the reviews say of
MALAYSIA'S ECONOMIC SUSTAINABILITY:
CONFRONTING NEW CHALLENGES
AMIDST GLOBAL REALITIES (2002)

"Written in a down-to-earth manner, it is a useful book to have. ...
essential reading for anyone keen to find out how Malaysia
succeeded in uplifting its economy while so many other Third
World countries are still struggling to develop, and what changes are
necessary if we are to continue moving forward in the face of
globalisation, internationalisation and intense competition."
New Straits Times

"Navaratnam is a sharp observer of Malaysian affairs as well as a
prolific writer. Combine these two qualities, and we have a trove of
good practical ideas which Malaysian leaders would do well to pay
heed to in the management of the country's political and economic
affairs." *The Star*

"... attempts to address some of [the socioeconomic] challenges by
providing thought-provoking arguments while analysing significant
economic policy developments that affect the lives of Malaysians. ...
Recommended for those who want to catch up on the
socioeconomic scene in Malaysia." *Malaysian Business*

"... The overwhelming slant in these commentaries is very refreshing;
nearly all of them address the question 'what next?' or 'how can we
do better?' With this approach, Navaratnam aims to stimulate
public debate on economic management and further build on the
economic success already achieved in Malaysia. [He] reminds us
constantly that the social and human angles are an integral part of
economic management. [He] does not subscribe to the sacred cows
of economic management and reminds the reader regularly to be
wary of standard recipes. ... It is a must for all analysts who wish to
understand Malaysia's economic success and the key debates and
issues which figure in its future development."
Robert Poldermans *DFC Benelux*

"Some may find ... his opinions to be provocative, but that is exactly what he strives to achieve, i.e., to stimulate public debate and provoke discussion on national economic matters at different levels of Malaysian society." *Management*

"... takes a no-nonsense look at how Malaysia has weathered the economic shocks of the past four years. ... [Navaratnam] deals in broad journalistic strokes that make the book direct, digestible and dippable. ... [He] is a Malaysian in the truest sense of the word, and this book serves to echo the concerns (and grouses) of the typical Malaysian." *Education Quarterly*

"... 34 hard-hitting essays through the eyes of a senior civil servant. An insightful analysis of the current state of play of the Malaysian economy. Navaratnam does not sidestep or relegate to the sidelines those Malaysia-specific sensitive issues—he boldly confronts them and offers pragmatic solutions. A worthy contribution for practitioners, policymakers and academics. A must-read for anyone seeking to understand the psyche of Malaysia in a global perspective." **Professor Bala Shanmugam** *Chair of Accounting and Finance, Monash University Malaysia*

"... Navaratnam uses straightforward prose to not only address the questions he asks, but also to give a board overview of the Malaysian economy, ..." *John Kennedy School of Government, Harvard University, Autumn 2002*

What
MALAYSIA'S SOCIOECONOMIC CHALLENGES: DEBATING PUBLIC POLICY ISSUES (2003)
is all about:

This book analyses Malaysia's ability to respond to the socio-economic challenges of globalisation. With the commencement of the Asean Free Trade Area (AFTA) in 2003, the movement towards the formation of Asean+3 (China, Japan and South Korea), and the prospects of some Asean countries establishing free-trade agreements with Australia, New Zealand and the United States, the challenges of greater competition will increase substantially. Furthermore, the rising influence of the Chinese economy will impact adversely on the economies of the region.

The haunting question is, Are we ready? Is Malaysia competitive enough to confront the challenges of globalisation? By commenting critically but constructively on Malaysia's public policy issues as they unfold, as well as on the nation's ability to respond to the threats and opportunities of globalisation, Navaratnam warns that Malaysia will fail to attain the goals of Vision 2020 if we do not enhance our competitive edge at a faster pace and improve on the affirmative-action policies.

ISBN 967 978 848 2

What the reviews say of
MALAYSIA'S SOCIOECONOMIC CHALLENGES: DEBATING PUBLIC POLICY ISSUES (2003)

"Navaratnam's writings offer a series of civilised dialogues. He responds to Prime Minister Mahathir Mohamad and numerous other Malaysian leaders; we also encounter some of the key international analysts, and a respectful debate with a Singapore high commissioner. He writes lucidly, moving comfortably between economics and politics, and making the foreign reader familiar with key debates taking place in Malaysia at a time when that country has attracted much international interest."
Anthony Milner *Basham Professor of Asian History, Dean of Asian Studies, Australian National University, Canberra*

"In this book, Navaratnam analyses a wide range of economic issues that confronts Malaysia today. Unlike many other commentators, his analysis is informed by deep insights acquired as an economist with the government for many years married to the wide experience he has gained as a corporate player in more recent times. This is why his views on the Malaysian economy deserve serious attention from both policymakers and captains of industry. Because his lucid articulation of the challenges facing the nation comes from the heart, the Malaysian public as a whole will also benefit tremendously from his book." **Dr Chandra Muzaffar** *President, International Movement for a Just World (JUST), November 7, 2003*

"This book is a must-read for all who are concerned about the balance of power in the global trading system. Many of our international institutions are almost 60 years old and have not been revisited to take into account the special circumstance of emerging markets. Navaratnam's book is a hard-hitting effort to bring these issues to the surface." **Gail Fosler** *The Conference Board, New York*